ENERGY IN WATER RESOURCE RECOVERY FACILITIES

WEF Manual of Practice No. 32

Second Edition

2021

Water Environment Federation
601 Wythe Street
Alexandria, VA 22314–1994 USA
http://www.wef.org

Cover photo courtesy of Pete Cavagnaro, Water Resources Recovery Solutions, LLC.

ISBN 978-1-57278-402-4

About WEF

The Water Environment Federation (WEF) is a not-for-profit technical and educational organization of 35,000 individual members and 75 affiliated Member Associations representing water quality professionals around the world. Since 1928, WEF and its members have protected public health and the environment. As a global water sector leader, our mission is to connect water professionals; enrich the expertise of water professionals; increase the awareness of the impact and value of water; and provide a platform for water sector innovation. To learn more, visit www.wef.org.

Prepared by the Energy in Water Resource Recovery Facilities Task Force of the Water Environment Federation

Peter V. Cavagnaro, PE, BCEE, PMP, *Chair*
Thomas E. Jenkins, PE, *Vice-Chair*

Mark K. Addison, PE
Shayla Allen
Sandra Angarita
Eric Auerbach
Hossain Azam, PhD, PE
Wendy Barrott
Lucas Botero, PE, BCEE, ENV SP
Alexandra Bowen, PE
Ken Brischke
Peter J. Bugg
Arturo Burbano
Marie S. Burbano, PhD, PE, BCEE
Joseph Cantwell
Tony Cecchini
Chein-Chi Chang
Mark Cusac, PE
Jens Dancer
Alexandra Doody, PE
John D. Dyson
Houssam El Jerdi, MS, PE
Christian Ference
Rachel Feil
Richard Finger
James Fischer
Robert (Bob) Forbes, PE
Josh Gable
Mark Greene
Georgine Grissop
Jim Groman
Mahmudul Hasan
Angela M. Hintz
Eric Hjellen
Jay Kemp
Ed Kobylinski
Chuck Kronk
Mark E. Lang
Xiaojin Li, PhD
Karthik R. Manchala
Lirane Mandjoupa
William Mattfeld
Bill McConnell, PE
Henryk Melcer
Jen Muir
Maureen Neville, PE
John Norton, PhD, PE
Ali Oskouie
Ned Paschke, PE
Vikram M. Pattarkine, PhD, BCEEM
Arifur Rahman, PhD, PE
Tanja Rauch-Williams
Eric Redmond
Matt Seib, PhD, PE
Andy Shaw, PhD, PE
Rajesh Shenoy
Alice Strazzabosco, PhD
Anthony Tartaglione, PE, BCEE
Will Thies
Peter Thomson
Donald Voigt
Jennifer Wagner
Andrea Turriciano White, PE
Rob Wiley
Ethan Willats
Hong Yin

Under the Direction of the Municipal Resource Recovery Design Subcommittee of the Technical Practice Committee

Manuals of Practice of the Water Environment Federation

The WEF Technical Practice Committee (formerly the Committee on Sewage and Industrial Wastes Practice of the Federation of Sewage and Industrial Wastes Associations) was created by the Federation Board of Control on October 11, 1941. The primary function of the Committee is to originate and produce, through appropriate subcommittees, special publications dealing with technical aspects of the broad interests of the Federation. These publications are intended to provide background information through a review of technical practices and detailed procedures that research and experience have shown to be functional and practical.

Water Environment Federation Technical Practice Committee Control Group

Andrew R. Shaw, PhD, PE, *Chair*
Jon Davis, *Vice-Chair*
Dan Medina, PhD, PE, *Past Chair*

H. Azam, PhD, PE
G. Baldwin, PE, BCEE
S. Basu, PhD, PE, BCEE, MBA
M. Beezhold
P. Block, PhD
C.-C. Chang, PhD, PE
R. Chavan, PhD, PE, PMP
M. DeVuono, PE, CPESC, LEED AP BD+C
N. Dons, PE
T. Dupuis, PE
T. Gellner, PE
C. Gish
V. Harshman
G. Heath, PE
M. Hines
M. Johnson
N.J.R. Kraakman, Ir., CEng.
J. Loudon
C. Maher
M. Mulcare
C. Muller, PhD, PE
T. Page-Bottorff
T. Rauch-Williams
V. Sundaram, PhD, PE
M. Tam, PE

Contents

List of Figures

List of Tables

Preface

This manual discusses principles and concepts of energy sources and uses in the transport, treatment, and recovery of water and water resources. The manual addresses utility billing procedures and incentives as well as energy considerations for electric motors and transformers, pumps, controls, aeration systems, blowers, energy generation, and solids processes.

Energy is often a major category of cost associated with wastewater treatment—this manual will provide insights that can reduce those costs by 10% to 50% or more. Additionally, revenues from resource recovery can provide a stream of income that will support other improvement activities. This second edition of the manual was produced under the direction of Peter V. Cavagnaro and Thomas E. Jenkins.

Authors' and reviewers' efforts were supported by the following organizations:

Aqua-Aerobic Systems, Inc.
Black & Veatch
Cambrian Innovation
CDM Smith
Centrisys
George Washington University
Great Lakes Water Authority
Gross-Wen Technologies
Hamlett Environmental Technologies
Jacobs
JenTech Inc.
Moleaer Inc.
Nuove Energie USA Inc.
Ovivo USA, LLC
PEACE USA
Pima County Regional Water Reclamation Department
University of the District of Columbia
Water Resources Recovery Solutions, LLC
Water Works Systems & Equipment, Inc
Xylem Inc.

1

Introduction

Marie S. Burbano, PhD, PE, BCEE;
Andrea Turriciano White, PE; & Christian Ference

1.0 INTRODUCTION

This manual is the *Energy in Water Resource Recovery Facilities*, Manual of Practice (MOP) 32, 2nd edition. The 2nd edition is an update to MOP 32, 1st edition, published in 2009. This 2nd edition continues its goal to be one of the principal references of contemporary practice for the energy-conscious design of water resource recovery facilities (WRRFs) and utilities of the future (UOTF). The purpose of the manual is to serve as a primer for helpful suggestions for optimizing energy and energy costs at a WRRF. It is intended for managers, facility engineers, and senior operators to provide a basis for greater understanding of the concepts of energy and efficient energy usage to better manage consumption and demand. It will also be useful to the design engineer in selecting energy-efficient equipment and processes. The second edition brings new focus areas with emphasis on the Water Resources UOTF, water resource recovery, and ongoing utility efforts and established practices to become net zero energy in terms of energy.

This manual discusses principles and concepts of energy requirements, potential inefficiencies, and recommended energy conservation measures (ECMs) for specific equipment and process used in the transport, treatment, and recovery of water resources.

The manual addresses utility billing procedures and incentives as well as energy considerations for electric motors and transformers, pumps, variable controls, aeration systems, blowers, energy generation, and solids handling processes.

This manual addresses energy and how it is used in the operation of municipal WRRFs. Information provided for pumping at WRRFs will also apply to raw water pumping stations, low- and high-service water pumps, water booster pumps, and wastewater lift stations found outside the WRRF fence line.

Energy is a major contributor of cost associated with wastewater treatment operations. Water and wastewater facilities are typically among the community's largest energy consumers (U.S. Environmental Protection Agency [U.S. EPA], 2013), accounting for 30% to 60% of a municipal government's energy usage (U.S. EPA, 2008) and 3% to 4% of the nation's total energy usage. Energy consumption is considered a significant operating cost at a WRRF, consuming 15% to 30% of the operation and maintenance budgets at a large WRRF and 30% to 40% at a small WRRF. Energy costs of operating these facilities continue to rise because of the trend in the cost of fuels, inflation, and increasing wastewater discharge requirements that result in the application of energy-intensive treatment processes.

Advanced technologies offer new opportunities for improving energy efficiency and the application of new energy conversation measures at a WRRF. Energy conservation measures are defined as a physical improvement, facility operation, or equipment maintenance practice that results in a reduction in energy usage or demand. Energy conservation measures are identified throughout this manual, as information about specific items of equipment and/or processes is discussed. This manual provides insights that can reduce those costs by 10% to 50% or more. Additionally, revenues from resource recovery can provide a stream of income that will support other improvement activities.

0=[p;vkThe manual includes basic concepts of energy and describes the technical basis for the energy required by unit operations and unit processes commonly found at WRRFs. A substantial amount of energy is also used in building- and facility-related equipment such as lighting and heating and ventilation. These subjects are already covered in references on energy conservation, some of which are listed in the "Suggested Reading" section at the end of this chapter.

2.0 WATER RESOURCE RECOVERY FACILITIES AND THE UTILITY OF THE FUTURE

The term *WRRF* is an evolution in the industry to consider wastewater as a resource with the potential to obtain valuable products like phosphorus and biogas using a variety of treatment technologies. As described in this manual, WRRFs are publicly owned facilities used to treat municipal wastewater, including domestic, commercial, institutional, and industrial waste streams that are discharged to the collection system.

Around the same time as the update to this manual, a new trend emerged and the concept of the UOTF was developed. There are several documents that have defined and established this new direction for the industry, including

- Water Environment Federation's (WEF) *The Energy Roadmap* (WEF, 2013),
- WEF's Policy and Position Statements, and
- NACWA/WEF/WERF *The Water Resource Utility of the Future: A Blueprint for Action* (NACWA et al., 2013).

Another concept that has been established by this new direction is the Nutrient-Energy-Water (N-E-W) paradigm. N-E-W has shifted the focus from energy efficiency (conservation and production) to the realization that energy is an overarching factor in the nature and treatment of wastewater. Wastewater contains materials (carbon) and characteristics (heat) that are significant sources of energy. Additionally, facilities with a focus on the N-E-W paradigm may "partner" with others in their geographic area to accept high-strength wastes for conversion to energy. There are several documents that further explain the N-E-W paradigm, including

- WEF's *The Nutrient Roadmap* (WEF, 2015),
- WEF's *The Water Reuse Roadmap* (WEF, 2018), and
- WEF's Technical Manual, *Moving Toward Resource Recovery Facilities* (WEF, 2014).

3.0 ENERGY TERMINOLOGY

This section provides definitions for energy terminology applied throughout this manual.

- Energy—the ability or capacity to do work
- Kinetic energy—the energy possessed by a moving object
- Potential energy—the energy held by an object because of its position relative to a zero position or other factors
- Power—the rate at which work is performed, measured as work per unit time

- Energy conservation measures—a physical improvement, facility operation, or equipment maintenance practice that reduces utility or operating cost
- Benchmarking—the comparison of a WRRF's or UOTF's overall energy usage to published typical uses of electricity, fuel, or chemicals at other WRRFs or UOTFs
- Measurement and verification—the process of accounting for the savings associated with each ECM
- Baseline measurements—the analysis of existing energy bills and operating data to identify the current level of consumption, peak energy usage, and costs for an existing WRRF, UOTF, process, or system
- Field measurement—energy measurements collected in the field to establish energy baselines or quantify improvements

3.1 An Overview of Energy and Power

The term *energy* has different interpretations. In this manual, *energy* is defined by the science of physics and is the ability or capacity to do work.

There are many different types of energy. These different types can be divided into two broad classifications: potential energy and kinetic energy. *Kinetic en*ergy is possessed by a moving object. *Potential energy* is stored and contained within fuel oil or in an elevated water tank or reservoir. Potential energy is converted to kinetic energy when released. Energy can be changed from one form to another. Common forms of energy are chemical, electrical, mechanical, thermal, radiant, and nuclear. Most people are familiar with the conversion of fossil fuel through burning (chemical energy) to form steam (thermal energy).

The most common form of energy usage in WRRFs is electric energy. Natural gas is also commonly used as fuel in boilers to provide comfort and process heat. Other forms of energy that may be used at a WRRF are fuel oil, propane, and steam. The various forms of energy common to WRRFs are detailed in Chapter 3.

Power is the rate at which work is performed. It is the speed at which the work is done or the rate at which energy is expended. The basic unit of power in the International Standard (metric) system is the watt, which gives rise to the production of energy at the rate of one joule per second (J/s). In the U.S. Customary system, the common units of measure for power are horsepower (1 hp = 550 ft-lb/sec) and British thermal units per hour (Btu/hr). Note that watt and horsepower do not appear to be rate units, but they are. All power units are measured as work per unit time.

The size of a motor or engine is defined not by the total amount of work to be done, but by the rate at which it can do work. Electric motor power is commonly measured in kilowatts (horsepower), boiler output is measured in watts (Btu/hr or boiler horsepower = 33 520 Btu/hr), and air-conditioner power is measured in watts (Btu/hr or tons of refrigeration = 12 000 Btu/hr).

The power applied multiplied by the time it is applied represents total energy consumption, which is expressed as joules (J) or kilojoules (kJ). Energy consumption is typically measured in joules (kilowatt-hours [kWh]) if consumed as electricity and in kilojoules (British thermal units [Btu]) if consumed as a fuel.

- Energy (joule) = power (kW) × time (hr)
- Energy (hp-hr) = power (hp) × time (hr)
- Energy (Btu) = power (Btu/hr) × time (hr)
- Energy (ft-lb) = power (ft-lb/sec) × time (sec)

Measured at standard temperature (15.56 °C [60 °F]) and pressure (101.56 kPa [14.73 psia]), natural gas contains approximately 37 000 kJ/m^3 (1000 Btu/cu ft), and billing units are typically 2.8 m^3 (100 cu ft) of gas, 28 m^3 (1000 cu ft or dekatherm), 105 000 kJ (100 000 Btu, otherwise known as *therms*), or million Btu (MMBTU).

Fuel oil characteristically contains 3.9×10^4 kJ/L (140 000 Btu/gal), and consumption is measured in liters (gallons). Table 1.1 provides conversion factors for changing the unit of energy consumption from one measurement method to the unit of a different measurement method. In addition, Appendix A provides U.S. customary to International Standard conversions for the units used in this manual.

TABLE 1.1. Conversion chart for various forms and units of energy

	Work		Heat		Electric	
Unit	ft-lb	kg-m	Btu	kcal	hp-hr	kWh
ft-lb	1	0.1383	1.286×10^{-3}	3.241×10^{-4}	5.050×10^{-7}	3.766×10^{-7}
kg-m	7.231	1	9.302×10^{-3}	2.343×10^{-3}	3.654×10^{-6}	2.724×10^{-6}
Btu	777.9	107.5	1	0.2520	3.927×10^{-4}	2.928×10^{-4}
kcal	3086	426.8	3.968	1	1.558×10^{-3}	1.162×10^{-3}
hp-hr	1.980×10^6	2.737×10^5	2547	641.7	1	0.7457
kWh	2.655×10^6	3.671×10^5	3415	860.5	1.341	1

Note: kWh × 2.78 × 10^{-7} = joules

4.0 MANUAL OF PRACTICE ORGANIZATION

Following is a brief overview of some of the major topics covered under each chapter of the manual.

Chapter 1—"Introduction" provides an overview of the manual and includes basic technical terminology describing the forms and sources of energy.

Chapter 2—"Energy Management" presents ways to control electrical demand such as how costs can be reduced through a better understanding of energy consumption, utility rate structures, and effective energy management. Ways to use alternate energy sources for greater overall economy are addressed. Examples are referenced to demonstrate how to compare differences in initial capital cost with operating cost changes because of differences in equipment efficiency.

Chapter 3—"Energy and Power" describes the forms of energy used at WRRFs, reviewing electricity, natural gas, gasoline, fuel oil, high purity oxygen, and digester gas. Heat recovery and onsite generation are also described.

Chapter 4—"An Overview of Energy Use at Water Resource Recovery Facilities" provides an overview of where energy is used or generated in a WRRF and defines key considerations for improving energy efficiency and achieving energy neutrality.

Chapter 5—"Pumping Systems" describes the key parameters that influence input power and energy usage in pumping processes for WRRFs and collection systems, and discusses numerous opportunities to reduce pumping power and energy usage.

Chapter 6—"Mixing" presents how to compare the energy used by different types of mixers, discusses the level of mixing needed in anaerobic and anoxic zones in biological nutrient removal systems, and mixing requirements for chemical addition.

Chapter 7—"Aeration Systems" describes how oxygen requirements are determined, types of aeration equipment available, factors affecting design and oxygen transfer efficiency, operational considerations, and opportunities for energy conservation.

Chapter 8—"Blowers" explains the importance of blowers to the total power consumed at a WRRF, considerations for blower applications, factors in the operation of positive-displacement blowers, operating principles of centrifugal blowers, and retrofit opportunities.

Chapter 9—"Process Controls" discusses approaches, components and systems that can be used to control wastewater treatment processes to optimize energy efficiency. These encompass overall process decisions, control strategies, and equipment including instrumentation and control valves.

The chapter focuses primarily on the aeration process, as the most energy-intensive process at WRRFs, and also addresses ultraviolet disinfection.

Chapter 10—"Digester Gas to Energy" is devoted entirely to anaerobic digestion systems at WRRFs and the production, treatment, and use of digester gas as a means of recovering energy from the anaerobic digestion process.

Chapter 11—"Renewable Energy" discusses the fundamentals and current advancement of renewable energy technologies in WRRFs including solar (photovoltaic), thermal, hydraulic, and wind energy.

Chapter 12—"Looking Forward" discusses new and emerging technologies and applications for WRRF energy management.

5.0 REFERENCES

National Association of Clean Water Agencies, Water Environment Research Foundation, & Water Environment Federation. (2013). *The water resources utility of the future: A blueprint for action.* https://www2.nacwa.org/images/stories/public/2013-01-31waterresourcesutilityofthefuture-final.pdf

U.S. Environmental Protection Agency. (2008). *Ensuring a sustainable future: An energy management guidebook for wastewater and water utilities.* U.S. Environmental Protection Agency, Office of Wastewater Management. GS-10F-0337M.

U.S. Environmental Protection Agency. (2013). *Energy efficiency in water and wastewater facilities.* U.S. Environmental Protection Agency, State and Local Climate and Energy Program (EPA–430-R-09-038).

Water Environment Federation. (2013). *The Energy roadmap.* ISBN 978-1-57278-273-0.

Water Environment Federation. (2014). *Moving toward resource recovery facilities.* ISBN 978-1-57278-303-4.

Water Environment Federation. (2015). *The nutrient roadmap.* ISBN 978-1-57278-314-0.

Water Environment Federation. (2018). *The water reuse roadmap.* ISBN 978-1-57278-339-3.

6.0 SUGGESTED READINGS

American Council for an Energy Efficient Economy. (2010). *Local technical assistance toolkit: Energy efficiency in water and wastewater facilities.*

American Water Works Association. (2016). *Energy management for water utilities.*

Dranetz BMI. (1991). *Dranetz field handbook for electrical energy management.*

Efficiency Valuation Organization. (2012). *International performance measurement and verification protocol, concepts and options for determining energy and water savings* (Vol. 1, April). EVO 10000-1.2012.

New York State Energy Research & Development Authority. (2010). *Water & wastewater energy management.*

Spellman, F. R. (2013). *Water & wastewater infrastructure: Energy efficiency and sustainability.* CRC Press. ISBN 9781138382213.

Tarllo, S., & Kohl, P. (2015). *A guide to net-zero energy solutions for water resource recovery facilities.* IWA Publishing. eISBN 9781780407685.

Toliyat, H. A., & Kliman, G. B. (2004). *Handbook of electric motors.* CRC Press, Taylor & Francis Group.

U.S. Department of Energy. (2008). *M&V guidelines—Measurement and verification for federal energy projects* (Version 3.0). U.S. Department of Energy, Office of Energy Efficiency and Renewable Energy, Federal Energy Management Program.

U.S. Environmental Protection Agency. (1973). *Electrical power consumption for municipal wastewater treatment* (EPA–R2/73-281).

U.S. Environmental Protection Agency. (1978). *Energy conservation in municipal wastewater treatment* (EPA-430/94-77-011).

U.S. Environmental Protection Agency. (2006). *Energy conservation—Wastewater management fact sheet.* U.S. Environmental Protection Agency, Office of Water (EPA 832-F-06-024).

U.S. Environmental Protection Agency. (2008). Energy star—Buildings & facilities web site.

U.S. Environmental Protection Agency. (2010). *Evaluation of energy conservation measures for wastewater treatment facilities* (EPA 832-R-10-005).

Wisconsin Focus on Energy. (2016). *Energy best practices guide: Water and wastewater industry.*

2

Energy Management

Joseph Cantwell; Alice Strazzabosco, PhD; & Donald Voigt

1.0 ROLE OF MANAGEMENT

1.1 Energy Efficiency Is a Utility Goal

The overarching responsibility of a water resource recovery facility (WRRF) is to protect public health and the environment by meeting discharge limits. Management (owners) should also set goals for energy efficiency and management. Energy efficiency is defined here as conserving or producing energy. Energy efficiency programs should define, monitor, execute, and implement the programs.

Management must commit to provide leadership, encouragement, resources, and access data to conserve and/or produce energy.

Knowledge of how energy is consumed at a facility will provide personnel with a valuable tool to address energy use and provide insight for budget control. In addition, by becoming energy efficient, a facility can be a leader in the community providing guidance to industries, demonstrating the value gained through energy efficiency, and free up operating budget for use in other areas.

1.2 Value of Being Energy Efficient

Water resource recovery facility personnel, in general, understand the physical and mechanical concepts of electricity, how it works, and the associated safety procedures that must be followed to work with it. However, to have an effective energy management program, personnel also need to be aware of how and where energy is used at their facility. Said awareness will provide treatment personnel with insight in how to become energy efficient and address budget control.

A good starting point is for WRRF staff to have access to monthly energy bills to learn the information provided on their energy bill. Details on understanding your energy bill are presented in Section 2.4.5 in this chapter and in *Energy Best Practices Guide: Water & Wastewater*

Industry—Appendix B—Understanding Your Electric Bill [Wisconsin Focus on Energy, 2020]).

1.3 Reactive and Proactive Management

1.3.1 Improve Efficiency of Existing Facilities

Energy efficiency may be in the form of conserving energy or producing it. This chapter focuses on conservation. Production is addressed in Chapters 10 and 11.

Energy management and improving energy efficiency is an ongoing effort. The most cost-effective way to reduce cost is, typically, to optimize energy use. There always are opportunities to reduce consumption, demand, or costs as the result of innovation in operating techniques and developments in equipment design. Equipment size, poorly selected, on the verge of failure, and/or not properly installed or maintained tend to be less efficient and use more energy.

A WRRF needs to develop a baseline and benchmark for their energy efficiency program. A baseline is the amount of energy that a WRRF is presently consuming per a predetermined unit. A benchmark represents the best value that your WRRF should be able to obtain with the facilities and unit treatment processes available. A benchmark is the goal toward which a WRRF should make efforts to reduce its energy consumption and demand to meet.

1.3.2 Integrate Energy Efficiency into New Designs

Presently, in the wastewater industry, energy efficiency is, traditionally, obtained through auditing existing facilities, identifying energy-efficient opportunities, and implementing them. This needs to change. Energy efficiency should begin by becoming an integral component of all design efforts. Design professionals should work with WRRF owners to incorporate energy efficiency into all designs, irrespective of the size of the WRRF. Further, the design professional should present life-cycle cost analysis to promote and support the energy-efficient design.

1.4 Life-Cycle Cost Analysis

Life-cycle cost analysis is a method of analyzing the cost (capital and operating costs) of a system, piece of equipment, or product over its useful life. The analysis includes the energy used and associated cost for the forecasted lifetime of the component in the assessment.

Energy is a significant cost of ownership, and when factored into the economic analysis, can justify the purchase of more expensive equipment that uses less energy.

A reference for life-cycle cost analysis is *Life Cycle Costing Manual for the Federal Energy Management Program* (NIST, 2020) and *Maximizing the Value of Investments Using Life Cycle Cost Analysis* (ASCE & Eno, 2014).

Other life-cycle cost analysis approaches are available through the Hydraulic Institute (2001), and *Energy Best Practices Guide: Water & Wastewater Industry*—Appendix C—Economic Evaluation Process (Wisconsin Focus on Energy, 2020).

1.5 Support Energy Efficiency Education

Implemented energy efficiency opportunities at a facility can result in a 20% to 30% reduction in overall energy consumption. Energy efficiency education of facility staff can be an invaluable investment for the community, irrespective of the size of the facility.

To guide energy efficiency awareness and education, Wisconsin Focus on Energy has developed the *Energy Best Practices Guide: Water & Wastewater Industry* (2020). The U.S. Environmental Protection Agency (U.S. EPA) provides insight on addressing energy efficiency. The New York State Energy Research & Development Authority (New York State Energy Research & Development Authority, 2010, 2019) has published similar publications.

1.6 Energy Efficiency Communication

Presenting and explaining energy and cost savings associated with recommended improvements at a WRRF to decision-makers almost always results in a decision to proceed with the program or project. The presentation of the cost savings should include a discussion on the utilization of life-cycle cost analysis. The life-cycle cost analysis information presented should include

- a simple description of the cost saving opportunity,
- an assessment of the financial effect of any needed proposed modification,
- a description of what is required to acquire the savings,
- an additional description of how the modifications will result in savings over the life of the project, and
- the basic payback time for the modifications.

It is also desirable to present examples in which the proposed modifications are installed at a similarly sized facility or where a configuration similar to that proposed has been used. The examples should include the savings being realized to illustrate yearly operating savings.

2.0 MANAGEMENT SYSTEMS

2.1 Energy Management Plan

Energy use represents a significant yearly expense at a WRRF. Cost control through energy management should be a high priority for all WRRF managers. Energy management should include the development of an energy management plan.

2.1.1 Energy Management Plan

An effective energy plan can be developed in nine steps. The nine steps, shown in Figure 2.1 and defined below, provide a systematic approach to support, identify, assess, implement, and continue energy management.

FIGURE 2.1. Basic steps in building an energy management program. Reprinted with permission from Wisconsin Focus on Energy—Wisconsin's Energy Efficiency and Renewable Energy Program.

Step 1— Establish organizational commitment. Energy efficiency efforts need to be empowered; therefore, support from administration and management is vital.

Step 2— Assemble and initiate an Energy Team. Assemble committed representatives from management, administration, accounting, finance, compliance, engineering, research, and operation and maintenance.

Step 3— Develop a baseline of the facility's energy use. This can be developed with metered and sub-metered data or developed from knowledge of the operation of the facilities. See Energy Best Practices Guide—Water & Wastewater Industry (Wisconsin Focus on Energy, 2020)—Appendix A—Baseline Energy Use and Key Performance Indicators (KPI).

Step 4— Develop profiles of energy usage for major equipment types. Identify amount (percent) of energy each major process at the facility is using.

Step 5— Identify and assess project opportunities. Identify and assess no- (or low-) cost operational modifications then progress into assessment of capital cost energy efficiency opportunities.

Step 6— Prioritize opportunities for implementation. With opportunities identified and assessed, identify which projects would meet the facility's goals responsive to financial capabilities.

Step 7— Develop an implementation plan. Develop a plan to implement the prioritized opportunities and then implement the plan to meet the program goals.

Step 8— Track and report progress. Monitor the implemented projects to identify their level of success and report the findings to reflect achieved energy savings.

Step 9— Continually update plan and achieve energy management goals. Energy efficiency is a continuous effort, so assess progress and adjust goals as efficiency goals are achieved.

2.2 Energy Management Strategy

Water resource recovery facility personnel need to become familiar with the energy billing methodology and how operating strategy affects energy charges. They would need to learn about the information provided on the monthly energy bill, beyond the "amount due" line. Terms such as kilowatt-hours [kWh], kilowatts [kW], on-peak, off-peak, and demand charges are just a few of the items that will enable facility personnel to understand how their actions affect the monthly charge for electricity. (Details on understanding your energy bill are presented in Section 2.4.5 in this chapter.)

2.3 Implementation Plan

Energy management planning can take on many different forms. A comprehensive and effective implementation plan should include the general and customized characteristics of the energy efficiency opportunity under consideration. Audits and studies of energy use should pinpoint opportunities of both high and inefficient energy use. Plans may include specifics of no- or low-cost opportunities and capital projects to recommend and fund, but they should also provide for continuous monitoring of energy use and for a means to see when new inefficiencies will arise. Continuous monitoring and assessment of new technologies that may provide improved energy efficiency also need to be incorporated so the current plan can be updated continuously.

2.4 Management Guidance and Support

2.4.1 Design Guidance

When utilities are considering the design of a project, the selected designer should use available energy efficiency best practices as guidance to ensure that the design will be energy efficient from startup.

Most states have codes that address the design of wastewater treatment processes, but do not reference the need to address energy efficiency.

The conscientious review of equipment selection is extremely important because many facilities never grow to their forecasted design conditions, thus, resulting in improvements that actually continuously waste energy. The following four recommendations are offered to help designers become aware of the effect that design decisions have on energy efficiency.

1. Associated with design procedures is the comparison of energy-efficient operational options along with the standard options. The comparison should conclude with life-cycle cost analyses to assess the cost of ownership.
2. Equipment should be selected to meet present day treatment needs effectively and energy efficiently, as well as be able to meet forecasted intermediate needs and the projected design conditions. Present standard design approaches need to be adjusted to include energy efficiency, so all designs are energy efficient from startup through their design life.
3. The designer should review the proposed improvements and assess if the proposed system can meet the treatment performance requirements while being energy efficient throughout the life of the project.

4. Equipment selections can also affect the energy efficiency of a facility. The treatment capability of the equipment selected is determined through the application of codes. The designer needs to assess what combination of equipment size selection will be energy efficient when the improvement initiates operation and when design conditions are met.

2.4.2 Monitoring

It is an old cliché: "What gets measured gets done." The message is clear: Measuring something gives you the data you need to make sure you achieve what you set out to do. And, in the specific case of processes such as wastewater treatment, specific measurement of multiple ongoing sub-process elements can be crucial to the optimization of facilities.

Wastewater characteristics such as flow, dissolved oxygen levels, temperature, energy, and performance parameters (biochemical oxygen demand [BOD], suspended solids, phosphorus, nitrogen, and pH, among these) should be monitored and recorded versus time. In addition, turbidity and settleability can also improve the base knowledge when making operational changes that may be recommended or implied by an energy audit. Mechanical operational data is also important to collect from running time meters, elapsed time meters, and wattmeters.

Continuous energy monitoring (and associated logging) of operational parameters and development of KPIs should be among the first priorities in an energy management program. Such logging and review will result in many discoveries related to energy efficiency. Peak demand charges, redundant motors, and base line energy costs will be discovered. From the continuous data, alarm "trip points" can be readily set to alert facility staff to make changes to avoid a wide variety of expensive excursions.

2.4.3 Metering

Metering instrumentation has evolved in parallel with automation. Microprocessor technology has revolutionized the capability to meter most parameters. From "strap-on" flowmeters to photosensitive dissolved oxygen and suspended solids meters to phosphorus and nitrogen inferential monitoring provides process knowledge and the ability to project and refine process operation. The implication is that sub-metering and overall energy consumption metering will be tracked in parallel. Consideration should be given to one-time sub-metering of power (watts) at motors and related electrical consuming processes in which more than 5 horsepower motors (or equivalent) are involved. Variable frequency drives (VFDs) nearly always provide watts/power as one of the monitored parameters in their programmable logic controller. Tap into that available data, wherever practical, and bring such data to a central location for assimilation and comparison.

Energy audits (see Section 6) may use recording wattmeters mounted on major motors and/or main sub-electrical panels throughout a WRRF. To the extent that audits demonstrate savings, the WRRF should consider installing permanent sub-metering, recording, and alarm devices at the most effective locations. Integrating this information into supervisory control and data acquisition (SCADA) systems and/or asset management systems becomes the "next step" toward providing real-time information to the operator.

2.4.4 Operation Guidance

Each utility should review and assess its operation strategy to assess if it is the most process- and energy-efficient strategy for the WRRF. Each WRRF is unique and needs to identify the best process operation to achieve its regulatory permit while at the same time identify the most energy-efficient procedures to follow to meet the process strategy. Consideration should be given to when the WRRF is being billed for on- and off-peak use (kWh) and demand (kW) charges so an energy-efficient operations schedule can be developed and followed to meet both criteria, regulatory permit compliance, and energy efficient operation.

2.4.5 Understand Your Energy Bill

The electric bill for typical WRRFs will usually provide important data on energy use and costs: total use for a given time period, on-peak energy use (kWh) and demand (kW), off-peak energy use (kWh) and demand (kW), date and time when on-peak and off-peak demand occurred, customer charge, demand charge, energy charge, power factor, and fuel surcharge.

The bill may include multiple tariffs based on special costs to the utility. These tariffs vary but nearly all have "peak" demand charges and "time-of-day" cost changes. It is a good idea to meet with the energy provider account manager and review your bill, tariffs, and rate schedule at least annually. Frequently, the provider will offer free advice regarding optimal tariffs and data for your specific bill.

An important request a WRRF should make is to request "interval" data, which commonly presents energy use historical kilowatt (kW) values in 15-minute intervals. These data can be used to familiarize WRRF personnel to how much and when energy is used and when and how demand charges are set. The WRRF can also request that the rates be reviewed based on historical consumption and compared to other rates that are available. Also, review the bill each month because billing errors can occur, incorrect tariffs applied, incorrect meter readings, and variable number of days in the billing period.

2.4.6 Use of Data

One of the first steps to energy management is creating an electrical budget using historical information. With the data collected, estimates of usage for the past measurement period are chronicled and compared with the electrical use recorded on the utility wattmeter(s).

There are various approaches to estimating electrical use in the absence of wattmeters on all pieces of equipment. For constant-speed equipment, it is often easiest to use runtime meters and power draw for each piece of equipment. For pumps and variable-load equipment, runtime meters are seldom of any use for electric power and, as such, other estimating methods must be used.

2.4.7 Optimize Operation of Facility

When the weekly or daily energy use for each process has been identified, it is time to review the operation of each process component individually and then aggregate each process to determine how it can be optimized. Then it is of value to review the operation schedule to identify what process can be modified, adjusted, or delayed in operation to better operate during off-peak (use and demand) time periods.

2.4.8 Demand-Side Management

Demand is defined as the rate of using electrical energy over a specified period—typically 15 or 30 minutes. Demand is the charge billed to a customer by the utility company to maintain the generation, transmission, substation, and distribution capacities needed to meet the maximum demand.

Demand reading is measured either in kilowatts or kilovolt-amperes (kVA). In billing on a KVA basis, all power (apparent power) used, including nonworking (reactive) power, is measured and billed, whereas billing on a kilowatt basis only includes the working (active) power.

Nonworking power is required to produce magnetic fields for any inductive load, such as motors or transformers. The ratio of apparent power to real power is called "Power Factor." A surcharge may be levied for a low power factor when using the kilowatt basis. In instances in which a WRRF is charged for a low power factor, it may make sense to install power factor correction capacitor(s).

Smaller demand capacity customers are typically charged based on kilowatts. Larger demand capacity customers are typically charged based on KVAs. The demand charge is designed such that the customer pays a fair share of the energy utility's fixed investment in production, transmission, and distribution equipment required to meet the customer's maximum requirements. Although the peak demand periods are typically of a short duration, just like peak loadings, the reserve capacity must be available.

There is a ratchet clause in some rate structures in which demand charges are based on the highest demand achieved over a period of the past 6 or 12 months, even if the present month peak demand is lower than the previous peak demand. Utility companies having problems meeting the system-demand requirements will generally devise the rate structures with a ratchet clause. In some instances, the ratchet clause applies only to a certain percentage of the previous peak demand. In some instances, peak demand is measured according to on-peak and off-peak hours. Methods of demand charge and rate calculation are unique to each power company and must be understood before a program aimed at peak demand reduction is initiated.

Figure 2.2 presents a graph of months versus energy consumption and peak demand for the La Crosse, Wisconsin, WRRF. The graph has a peak demand occurrence range of 810 to 884 kW for the 12 months graphed. Consumption has a range of 440 000 to 590 000 kWh. Note that the consumption and demand values vary, and maximum consumption and peak demands do not occur during the same months. Plots of previous years of electric consumption that are like this should be reviewed to identify how the WRRF has been operating and to identify minimum demand and consumption targets.

In predominantly peak-causing billing, in which frequent simultaneous starting and stopping of equipment occurs in lieu of continuous running of equipment, the demand charge portion of the bill could easily exceed 50% of the total charges. Thus, demand management through the use of soft starts, VFDs, start controls, demand monitoring, and alarms should be considered.

2.4.9 Energy Efficient from Start to Finish

A WRRF should be designed and equipment selected to provide the WRRF personnel with a system that can be operated efficiently from startup through the useful design life of the equipment. Equipment selection needs to address code required design conditions. However, when the size or capacity of equipment is selected, the existing conditions, present low, average, and high loading conditions should be considered in the sizing selection so the operations personnel have properly sized equipment to operate energy efficiently at startup and still meet design loading forecasts.

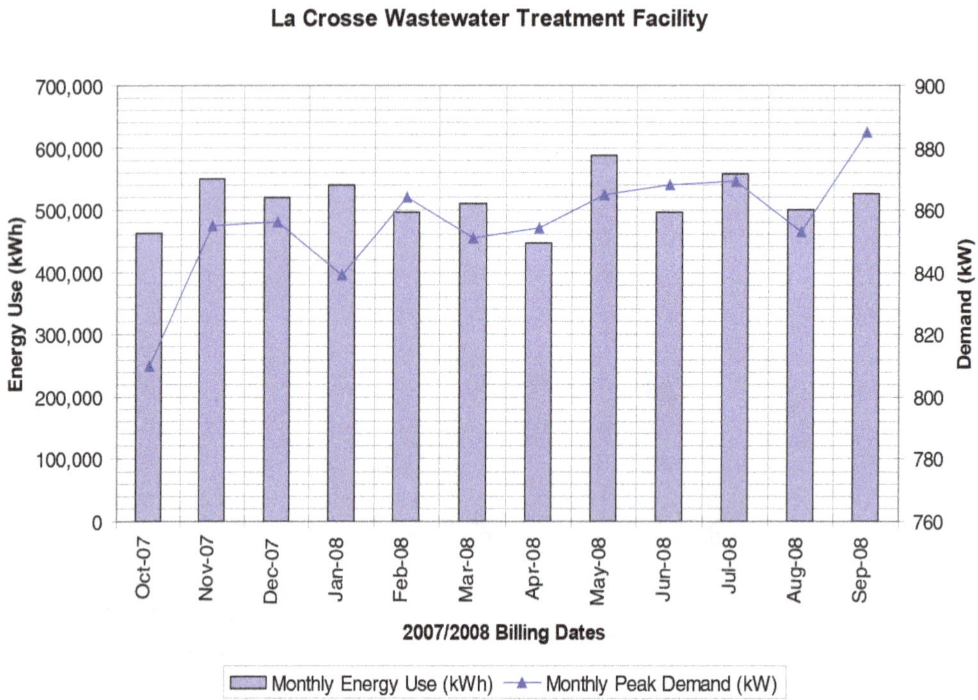

FIGURE 2.2. Months versus peak demand at a wastewater treatment facility (courtesy of Leidos Engineering, LLC). Reprinted with permission from City of La Crosse, Wisconsin.

3.0 IDENTIFING AND RANKING OF ENERGY CONSERVATION MEASURES

3.1 Define an Energy Conservation Measure

Energy conservation measures (ECMs) are defined as a physical improvement, facility operation, or equipment maintenance practice that results in a reduction in utility or operating cost. Energy conservation measures are evaluated and implemented with the aim to reduce energy use and cost for WRRFs. Energy conservation measures can be adopted in every aspect of a WRRF in which energy is consumed, including but not limited to pumping, aeration blowers, heating, cooling, and lighting. Energy conservation measures can be implemented one at the time or in multiples. However, to achieve the best outcome for the entire facility, ECMs should be systematically evaluated as part of a comprehensive energy improvement program. Energy conservation measures should align with existing objectives, plans, and programs of an existing facility.

Although ECMs are focused at an overall increase of energy efficiency, because of the way some electric bills are structured, it is not a guarantee that an increase of energy efficiency will translate into a cost saving. Careful evaluation of the overall energy consumption and demand, time of use, and ultimate bill effect associated with the ECM should be accounted for when planning energy management improvements to achieve energy efficiency in a cost-effective manner.

3.2 Describing an Assessment Approach

Although in new facilities energy efficiency can be incorporated into the design phase, increasing energy efficiency in existing facilities requires a clear identification of possible ECMs that can be beneficial and implemented.

3.2.1 Identification of Energy Conservation Measures

Identification of ECMs should begin by compiling a list of all the energy-consuming equipment. Units should be listed by unit process, typically in order of treatment processes at the WRRF site. The amount of time each unit is operated should also be recorded or estimated to the best ability of operations personnel. If data are not readily available, a performance test can be carried out to understand power draw and efficiency of the equipment. This evaluation should include equipment directly related to wastewater treatment and to equipment that consumes energy, including rooftop mounted heating, ventilation, and air conditioning (HVAC) units and lighting. Identifying and implementing ECMs is the fifth, sixth, and seventh steps of the nine-step Basic Steps in Building an Energy Management Program, described in Section 2.1.1.

3.2.2 Ranking, Evaluation, and Selection of Energy Conservation Measures

Once ECMs have been identified, it is important to assess them to establish which ECMs should be developed first. An ECM ranking matrix should be developed with consideration given to several factors, including costs of implementation, time to implement, future costs associated with operation and maintenance, the return on investment, and ease of implementation. Based on these factors, ECMs can be grouped in three categories: no or low cost and time (quick to implement); moderate cost and time, with some technical analysis required; and high cost, time, and complex engineering analysis required. This ranking process can be assisted by the adoption of analytical tools like the multicriteria decision analysis and decision matrixes.

Energy conservation measures are the first category (low cost, time, and quick to implement); they should be implemented first and they typically include changes in operational routines or the substitution of minor equipment (lighting). Cost associated with these measures are typically covered by the existing budget of the WRRF under exam. These ECMs should be implemented at the earliest possible date.

Energy conservation measures in the second category (moderate cost and time with some technical analysis required) may require staff evaluations, minor drawings or specifications, outside contractor assistance, and budget item procurement of labor and material. Energy conservation measures in this group can be compared and prioritized based on payback periods. The payback period is calculated dividing the cost of implementing the ECM by the net annual savings it will provide. Energy conservation measures with the shortest payback period are typically given the highest priority.

The last category (high cost, time with complex engineering, analysis required) of ECMs require careful evaluation by the engineering staff or consultants. Major structural rehabilitation, automation systems, major equipment change out, and changes to the treatment process are examples of ECMs in this group. Economic analyses that take into account the time value of money is required to compare alternative high-cost ECMs and justify their implementation.

3.2.3 Estimate of Energy Reduction

Estimating the energy savings from energy efficiency programs is a critical evaluation step to calculate the benefits associated with the ECMs. The effect of the energy efficiency programs is calculated taking the difference between

- *predicted* energy consumption after efficiency measures are installed and
- the estimate of what energy consumption would have been during the same period if the efficiency measures had not been installed (i.e., the adjusted baseline).

The baseline and the *predicted* energy consumption should be adjusted for factors affecting energy consumption, such as weather, seasonality of operations, and operating hours.

Energy savings are therefore determined using the equation:

Energy savings = Baseline energy use − Post-installation potential energy use ± Adjustments

3.2.4 Implementation Cost

In its simplest form, implementation costs include the cost of replacement of physical equipment with a more efficient version. For the more extensive ECMs, additional costs must be considered, such as the cost of planning, design, construction, and financing. A life-cycle analysis for all the costs associated with the project development should be carried out for each ECM considered.

3.2.5 Forecast of Payback Period

The payback period of an ECM measures how long it takes for the implemented measure to break even on the cost of the initial investment and is defined as: Simple Payback period = Cost of project ($) / Annual energy savings ($/year). The payback period depends on initial costs, ongoing costs (i.e., operating and maintenance), variation in price of energy, and the ECM saving as a result of the reduced energy purchase. When performing a payback analysis, the variation in annual energy savings needs to be estimated in facilities where growth is accounted for. Together with the initial cost of the investment, the payback period represents the selection criteria used to decide on the implementation of ECMs. An example is provided in *Energy Best Practices—Water & Wastewater Industry*—Appendix C (Wisconsin Focus on Energy, 2020).

Forecasting the return on investment is highly influenced by fluctuation of the energy price. Highly unstable prices may represent a barrier to the development of ECMs because of the effect on payback analysis.

4.0 ENERGY BASELINES

4.1 Describing and Defining Baseline

An energy baseline is the reference point that allows determination of project savings or project performance by comparison to post-project energy performance. The baseline is measured over a fixed time period—commonly, a 12-month time period. It may also be a calendar year, fiscal year, other appropriate 12-month period, or another period as determined appropriate. The baseline represents the energy consumption of the facility that serves as the basis of comparison to subsequent performance. A more accurate approach to establish the energy performance of a WRRF is to establish energy baselines for individual energy consuming equipment, unit treatment processes, or the entire WRRF. However, this is more costly and time-consuming approach. It can identify the level of energy demand and peak energy usage during a representative period of energy consumption or identify when and where energy consumption occurs. The energy performance of a WRRF should be periodically checked against the baseline to assess if energy consumption follows the anticipated trends.

Establishing the baseline is the first action to take before attempting to reduce energy consumption. The effect of each ECM can be measured against it. A baseline is drawn using the energy data of existing energy bills and measured data. Once ECMs are installed, and the savings (additional cost) determined, the baseline should be adjusted with the new energy consumption values.

4.2 Identifying a Baseline for Your Facility

Once energy data are gathered, KPIs should be defined to establish one or more baselines and to measure the performance of the facility. Several types of KPIs have been identified for WRRFs. A commonly adopted methodology is to use a KPI that represents the entire energy consumption of the facility (kWh/year, kWh/day, kWh/population equivalence [PE], and kWh/MG). These types of KPIs are useful when detailed energy data are scarce; however, they provide no indication on the energy performance of each stage of the treatment processes.

A more accurate methodology suggests defining KPIs for each stage of the wastewater treatment process as proposed by the European Commission ENERWATER (2015). These KPIs are, therefore, based on the key pollutants removed and enable one to draw a more detailed picture of the energy consumption of a WRRF, as follows:

- kWh/lb NH_4-N removed,
- kWh/lb BOD removed,
- kWh/lb chemical oxygen demand removed,
- kWh/lb of total suspended solids removed, and
- kWh/lb of total nitrogen removed.

Another KPI to consider is kWh/PE. This KPI could provide better comparison of WRRFs that have a high industrial contribution when seeking insight on process treatment efficiencies. The challenge is there is not a global value for a PE.

Defining multiple baselines can be costly and time consuming. It is, therefore, important to identify the best KPI based on energy data availability and accuracy, the characteristics of the WRRF (size, discharge requirements, flow fluctuations, etc.), and the energy efficiency targets. A methodology developed by the EPA Ireland proposes a two-step approach to the identification of KPIs and the development of energy efficiency baselines. In the first step, operators complete an assessment of the type of energy data available (kWh/lb BOD removed, kWh/lb total nitrogen removed) and their accuracy and consistency over time. Once it is clear what data are available, it is possible to compile a thorough list of all the available KPIs and identify the necessary ones (refer to the EPA [Ireland] Research Report 168 [2016]).

5.0 ENERGY BENCHMARKING

5.1 Describing and Defining Benchmark

Benchmarking is the practice of comparing the measured performance of a device, process, facility, or organization to itself, its peers, or established norms, with the goal of informing and motivating performance improvement. U.S. EPA has established programs to assist WRRFs in establishing benchmarks for their facilities. Energy Star Portfolio Manager provides a methodical approach (specific to wastewater facilities) allowing the user to access and calculate benchmark consumption.

5.2 Determining the Benchmark for Your Water Resource Recovery Facility

Benchmarks (for energy) are developed for an individual WRRF. The value should reflect the minimum amount of energy the WRRF's unit treatment processes need to treat the influent to meet its effluent limits. Once this value is established, it should become the energy consumption goal for the WRRF. Comparative benchmarking data is available at a number of locations, including *Energy Best Practices Guide—Water & Wastewater Industry* (Table 3: Average Energy Use at Wisconsin Wastewater Facilities) (Wisconsin Focus on Energy, 2020).

Once a WRRF is benchmarked, a routine tracking (monthly and annually) graph will be helpful to all WRRF staff to evaluate progress in improving WRRF performance against its own benchmark as well as comparing to local, regional, and national values for efficiency.

To develop a target, it is first necessary to identify where you are today (i.e., develop your current energy baseline) and then assess and determine where you want to be tomorrow (i.e., develop your target benchmark).

This process will begin with conceptually dividing the WRRF into systems and then into process units, with the final breakdown to each piece of equipment. Water resource recovery facility efficiency identifies the array of interdependency that exists, as well as the level of interconnection among all the components. This process will lead to the best identification of savings. This will also provide operators with a baseline on the energy consumption per treatment system at the WRRF.

6.0 ENERGY AUDITS

An audit will identify and assess opportunities available to improve energy use and identify a level of energy use and cost reduction.

6.1 Why I Need an Energy Audit

Just as benchmarking is helpful in establishing comparable energy use values for similar WRRFs, an energy audit provides a clear picture as to your individual WRRF and how it is currently performing.

6.2 Describing and Defining Energy Audit

Energy audits should involve measurements and recordings of actual field conditions at the WRRF being examined. This data should include energy use, time of use, and magnitude of use. Wastewater quantity and quality data is also needed to use in developing the audit. This information is necessary so the audit can identify site-specific KPIs (as referenced in Section 4 of this chapter) for the WRRF.

6.3 Describing Levels of Energy Audits

The level of detail and degree of complexity of an audit will vary. The audit detail will vary responsive to the size of the WRRF, the unit treatment processes on site, the extent and level of detail of available data, and the sites' purpose to do the audit. A reference link to describe wastewater systems' audit approach is presented by U.S. EPA Pacific Southwest Region 9 https://19january2017snapshot.epa.gov/www3/region9/waterinfrastructure/audit.html #diy. Further detail of energy auditing and its approach, which is referenced by Region 9, is defined through American Society of Heating, Refrigerating and Air-Conditioning Engineers (ASHRAE) as Levels I, II, and III.

These levels, I, II, and III, correspond to respective levels of detail and measurement which may be useful when deciding on level of detail being sought by your energy audit, refer to ASHRAE 2011 at https://www.ashrae.org/.

Increasing energy audit levels will have increasing costs. As in the selection of any technical group for your WRRF, it is prudent to request proposals from prospective auditors with qualification portfolios based on a clearly defined request for proposals. Qualified organizations may include energy service companies and design/build firms, as well as independent consulting firms. Energy service companies develop, design, build, and assist with funding projects that save energy, reduce energy costs, and decrease operations and maintenance costs at their customers' facilities.

An excellent resource for levels of accreditation and relative competency is the Association of Energy Engineers (www.aeecenter.org). Credentials for higher level audits will normally encompass "licensed Professional Engineers" (PE) and Certified Energy Managers (CEM).

7.0 COMPUTER MODELS FOR DEMAND-SIDE MANAGEMENT

Computer models can be used to guide energy strategies for energy optimization, reduction, and integration of on-site renewable energy technologies (for new designs or for retrofitting existing facilities), and to monitor energy usage. Energy models are mostly built around a mathematical optimization framework and incorporate either economic theories and/or systems engineering principles. Feeding the model with information about the goal to achieve (e.g., cost reduction target, carbon neutrality, energy efficiency level) and technical specifications of ECM options and/or renewable energy sources, the model will enable to compare technology options, energy consumption, pollutant emissions, and marginal costs for energy resources, over time and by region.

7.1 Use of Models During Planning and Design

Energy models for planning and design have the main purpose of aiding in the analysis of the energy, environmental, and economic effect of various energy strategies. Comparing energy options enables a designer to identify the "optimum" solution based on the predefined conditions and set goals. Care should be taken to match the mixing design parameters included in the model with the actual parameters. For example, the mixer energy in a basin default value in the model could be say 40 kW/1000 cu ft of basin volume—where, in reality, the installed mixer horsepower could be either high or low.

A list of commercial software developed to model WRRFs that include modeling of the energy usage is presented below.

- GPS-X (Hydromantis) provides a WRRF schematic output summary of the amount of energy used or the operating cost of the WRRF https://www.hydromantis.com/GPSXfeatures.html#energy_operating

- BioWin 5.0 (EnviroSim): BioWin 5.0 enables users to capture a facility-wide inventory of power demand, including blower power, mixing power, mechanical power, pumping power, aeration power, solid/liquid separation/disinfection power, and HVAC power. This allows the power requirements to be easily displayed and itemized. https://envirosim.com/3020
- The Carbon Heat Energy Analysis Plant Evaluation Tool (CHEApet) funded by Water Environment Research Federation is a static predictive modeling tool for quantifying and managing energy consumption and greenhouse gas estimation associated with wastewater treatment processes. https://www.accesswater.org/?id=-297270&fromsearch=true#iosfirsthighlight

7.2 Load Management

Monitoring energy consumption is an important step to identify areas of potential improvement and any atypical consumption. It reduces billing errors, waste, improves budgeting and operating practices, and identifies energy inefficiency. The U.S. EPA's online Energy Star for WRRF Portfolio Manager (see references) is an energy management and tracking tool that enables users to measure and track the energy and water use performance of a WRRF over time. A comprehensive list of energy management tools for WRRFs is available from the U.S. Department of Energy (DOE, n.d.).

7.3 Use of Models for Demand-Side Management

Demand-side management includes a set of actions that reduces on-peak demand and it differs from energy efficiency strategies. Energy efficiency actions reduce energy use (kWh) on-peak demand (kW) management does not necessarily reduce the peak, but manages it by shifting to off-peak rates, reducing costs. It can be done automatically or manually, in response to the energy demand signal, as illustrated in Section 2.4.8. Computer models can aid the automation of load management. In most cases, models used for energy monitoring serve more than one purpose and incorporate load management functions. The demand response Research Center (Berkeley Lab) for example, suggests using the existing centralized control system (SCADA) of a WRRF to coordinate and schedule energy load shedding and/or shifting through equipment-level control (Aghajanzadeh et al., 2015).

8.0 FINANCING APPROACHES

Fiduciary responsibility of WRRF management requires an identified project to be financially justified. Financial justification is typically through a simple linear annual savings payback approach. However, a more financially responsible approach would be using life-cycle cost analysis (or net present value analysis).

Both of these methods incorporate the beneficial value of efficiency into the analysis. Examples of both of these analysis are presented in *Energy Best Practices Guide: Water & Wastewater Industry*—Appendix C—Economic Evaluation Process (Wisconsin Focus on Energy, 2020).

After a project has been financially justified, a method to finance the project needs to be identified. Financing alternatives and options can be found at https://www.epa.gov/waterfinancecenter. This site presents

- effective funding frameworks,
- leading edge financing,
- financing resilient and sustainable infrastructure, and
- finance webinars and forums.

Additional financial information is located at https://www.dsireusa.org/, a database of state incentives for renewables and efficiency (DSIRE, 2020). Use this comprehensive database to find city, state, federal, and utility incentives and policies related to energy efficiency and renewable technologies. The database includes financial incentives such as tax incentives, rebates, grants, loans, and more.

9.0 REFERENCES

Aghajanzadeh, A., Wray, C., & McKane, A. (2015, August). *Opportunities for automated demand response in California wastewater treatment facilities*. Berkeley Lab. https://www.osti.gov/servlets/purl/1233609

American Society of Civil Engineers & Eno Center for Transportation. (2014). *Maximizing the value of investments using life cycle cost analysis*. American Society of Civil Engineers. https://www.asce.org/uploadedFiles/Issues_and_Advocacy/Our_Initiatives/Infrastructure/Content_Pieces/asce-eno-life-cycle-report.pdf

DSIRE. (2020). Database of state incentives for renewables and efficiency. http://www.dsireusa.org/

ENERWATER. (2015). *Standard method and online tool for assessing and improving the energy efficiency of waste water treatment plants*. Identification of key parameters and key performance indicators (KPIs) and ENERWATER methodology. http://www.enerwater.eu/wp-content/uploads/2015/10/ENERWATER-D3.1-ENERWATER-Methodology-v0.pdf

Environmental Protection Agency (Ireland). (2016). *Increasing resource efficiency in wastewater treatment plants* (Research report 168). http://www.epa.ie/researchandeducation/research/researchpublications/researchreports/researchreport168.html

Hydraulic Institute. (2001). *Pump life cycle costs: A guide to LCC analysis for pumping systems*. https://www.energy.gov/sites/prod/files/2014/05/f16/pumplcc_1001.pdf

National Institute of Standards and Technology. (2020). Life cycle costing manual for the federal energy management program. U.S. Department of Commerce. https://nvlpubs.nist.gov/nistpubs/hb/2020/NIST.HB.135-2020.pdf

New York State Energy Research & Development Authority. (2010). *Water & wastewater energy management, best practices handbook*. https://www.nyserda.ny.gov/-/media/Files/Programs/Clean-Energy-Communities/NYSERDA-Water-Wastewater-Energy-Management-Best-Practices-Handbook.pdf

New York State Energy Research & Development Authority. (2019). *Wastewater energy management, best practices handbook*. https://www.nyserda.ny.gov/-/media/Files/Programs/SEM/Best-Practices-Guide-Wastewater-Energy-Management.pdf

U.S. Department of Energy Better Buildings. (n.d.). The website provides a list of specific tools available to identify opportunities, monitor progress, and improve efficiency in any WWTF. https://betterbuildingssolutioncenter.energy.gov/better-plants/software-tools

Wisconsin Focus on Energy. (2020). *Energy best practices guide—Water & wastewater industry*. https://focusonenergy.com/sites/default/files/WW-Best-Practices_web.pdf?_ga=2.183994914.540090790.1528053189-1887821356.1525871497

10.0 SUGGESTED READINGS

Australian Government, Office of Environmental Heritage. (2019). *Energy Efficiency Opportunities in Wastewater Treatment Facilities*. https://www.environment.nsw.gov.au/resources/business/wastewater-treatment-facilities-energy-efficiency-opportunities-190114.pdf

Cantwell, J. (2010). Overview of state energy reduction programs and guidelines for the wastewater sector, intelligence online. https://iwaponline.com/ebooks/book/538/Overview-of-State-Energy-Reduction-Programs-and

Consortium for Energy Efficiency. (n.d.). *CEE national municipal water and wastewater facility initiative.* https://library.cee1.org/system/files/library/2650/ww-init-des.pdf

Crawford, G. (2010). *Technology roadmap for sustainable wastewater treatment plants in a carbon-constrained world.* IWA Publishing. https://www.iwapublishing.com/books/9781843393962/technology-road-map-sustainablewastewater-treatment-plants-carbonconstrained?utm_source=TrendMD&utm_medium=cpc&utm_campaign=IWA_Publishing_TrendMD_0

Crawford, G., & Sandino, J. (2010). *Energy efficiency in wastewater treatment in North America: A compendium of best practices and case studies of novel approaches.* https://iwaponline.com/ebooks/book/536/Energy-Efficiency-in-Wastewater-Treatment-in-North

Daw, J., Hallet, K., DeWolfe, J., & Venner, I. (2012) *Energy efficiency strategies for municipal wastewater treatment facilities.* osti.gov. https://www.osti.gov/biblio/1036045

Fuller, S. K. (2005). *Guidance on life-cycle analysis required by executive order 13123.* Prepared for the Department of Energy Federal Management Program, Washington, D.C., by S. K. Fuller, Economist, Office of Applied Economics, Building and Fire Research Laboratory, National Institute of Standards and Technology, Gaithersburg, MD 20878. https://www.energy.gov/sites/prod/files/2013/10/f3/lcc_guide_05.pdf

Greenberg, E. *Energy audits for water and wastewater treatment plants and pump stations.* CED Engineering. https://www.cedengineering.com/userfiles/Energy%20Audits%20for%20Water%20&%20Wastewater.pdf

Hernandez, V., Bellver-Domingo, A., & Hernandez-Sancho, F. (2018). Efficiency of wastewater treatment facilities: The influence of scale economies. *Journal of Environmental Management, 28*, 15 December 2018, 77–84. https://www.sciencedirect.com/science/article/pii/S0301479718310077

ISO. (2018), ISO 50001. Energy Data Management Systems. https://www.iso.org/iso-50001-energy-management.html

Long, S., & Cudney, E. (2012). *Integration of energy and environmental systems in wastewater treatment plants.* ieefoundation.org. http://www.ieefoundation.org/ijee/vol3/issue4/IJEE_03_v3n4.pdf

National Association of Energy Service Companies. (2020). Washington, D.C. http://www.naesco.org; Energy Services Coalition (ESC). https://www.energyservicescoalition.org/

Sebastian, J., Kumar, P., Hegde, K., Brar, S., Verma, M., & Surampalli, R. (2020). *Sustainability and energy management in facilities for wastewater treatment and reuse*. Wiley & Sons. https://doi.org/10.1002/9781119434016.ch26

U.S. Department of Energy Better Buildings. (2017). *Energy data management manual for the wastewater treatment sector*. https://www.energy.gov/sites/prod/files/2018/03/f49/WastewaterTreatmentDataGuide_0122.pdf

U.S. Department of Energy. (2017). *Energy data management manual for the wastewater treatment sector.* https://www.energy.gov/sites/prod/files/2018/01/f46/WastewaterTreatmentDataGuide_Final_011 8.pdf

U.S. Department of Energy. (2020). *Federal energy management program*. https://www.energy.gov/eere/femp/utility-energy-service-contracts-federal-agencies

U.S. Environmental Protection Agency. (2008, January). *Ensuring a sustainable future: An energy management guidebook for wastewater and water utilities*. https://www3.epa.gov/npdes/pubs/pretreatment_ensuring_sustainable_future.pdf

U.S. Environmental Protection Agency. (2010a). *EPA's energy use assessment tool user's guide* (version 2.0). https://www.epa.gov/sites/production/files/201601/documents/energy-use-assessment-tool-user-s-guide.pdf

U.S. Environmental Protection Agency. (2010b). *Evaluation of energy conservation measures for wastewater treatment facilities*. https://nepis.epa.gov/Exe/ZyPURL.cgi?Dockey=P1008SBM.TXT

U.S. Environmental Protection Agency. (2016). *Evaluation of energy conservation measures for wastewater treatment facilities*. https://www.epa.gov/sites/production/files/2016-01/documents/p1008sbm.pdf

Water Environment Federation. (2013). *The energy roadmap*. https://news.wef.org/wef-energy-roadmap-serves-as-user-guide-for-utilities-of-all-sizes/

Xingzhou, Y., Wei, J., Ye, G., Zhao, Y., Li, Z., Qiu, G., Li, F., & Wei, C. (2020). The correlations among wastewater internal energy, energy consumption and energy recovery/production potentials in wastewater treatment plant: An assessment of the energy balance. *Science of the Total Environment*, *714*, 20 April 2020, 136655. https://www.sciencedirect.com/journal/science-of-the-total-environment

3

Energy and Power

Jen Muir & Marie S. Burbano, PhD, PE, BCEE

1.0 FORMS OF ENERGY USED AT WATER RESOURCE RECOVERY FACILITIES

The most common form of energy used in wastewater facilities is electric energy. Natural gas is commonly used as a fuel in boilers for generation of heat and electricity. Other forms of energy that are used at a treatment facility site are fuel oil, propane, and steam. Programs are now available that allow for the direct purchase of power from renewable "green" energy sources, such as wind and solar power, under long-term contracts with fixed rates. As the cost of power from conventional sources continues to rise, these fixed-rate contracts from green sources will eventually produce savings and reduce the carbon footprint at water resource recovery facilities (WRRFs).

2.0 WASTEWATER

2.1 Thermal Content

The thermal content of wastewater is frequently higher than that of potable water because of the heat generated from showers, clothing washing machines, cooking, and human waste. Thermal content can be defined by the following equation:

$$E = m \times h(t)$$

Where E is the energy content in joules/hr (Btu/hr), m represents the mass flowrate of the wastewater in kg/hr (lb/hr), and $h(t)$ is the enthalpy of the stream in joules/kg (Btu/lb).

Wastewater temperatures typically have some seasonal variability, and many U.S. facilities have an average of 16 °C (60 °F) throughout the year. This thermal consistency is necessary for parts of the treatment process; however, the excess thermal energy presents the opportunity to provide both supplemental heat during the winter months and heat dissipation in the summer months through the usage of heat pumps.

2.1.1 Heat Recovery

As the cost of energy continues to rise, more and more facilities are examining waste heat recovery and figuring out ways to incorporate it into the WRRF or on a larger scale throughout a community. Waste heat recovery can be used to heat buildings or processes at the WRRF and/or other buildings located in close proximity to wastewater piping. The operation of a thermal heat pump is similar in concept to a refrigeration system. Heat pumps use a heating source and transfer this heat energy to a cold source. As the cold source is heated by the pump, the energy from the heating source dissipates and becomes cooler. Heat pump performance can be defined by the coefficient of performance (COP), as shown in the following equation:

$$COP = Q/W$$

Where Q is the heat output of the pump and W is the work input of the pump. Heat pumps typically have a COP value of 3 or 4 and can range from 2 to 5. The efficiency of heat pumps is dependent on multiple factors including the waste heat temperature, the desired temperature of the space being heated, and the energy cost of the energy used to operate the heat pump. Determining an approach to make a heat recovery system economically beneficial is currently the biggest hurdle to waste heat recovery systems, but improved technology and widespread funding programs have made these systems more desirable in recent years.

2.2 Organic Content

The most commonly used energy resource in wastewater is the organic fraction, often measured as the chemical oxygen demand and biochemical oxygen demand (e.g., BOD_5, that is, 5-day BOD measurement), which indicates the amount of oxygen (O_2) required to oxidize the organic and inorganic material to carbon dioxide (CO_2) and water (H_2O).

The organic fraction is categorized into dissolved and suspended. Organic fraction is also classified into biodegradable and refractory. The thickened part of organic sludge can be anaerobically digested for methane (CH_4) production. However, the methane production will only result from the biodegradable fraction of the sludge.

The organic content of the sludge can be accessed through various processes, including anaerobic digestion, that allow for additional methane production.

The dissolved organic fraction will not accumulate in the sludge as readily and typically go on to be treated by the aerobic treatment process, which can require significant energy input.

Maximizing the energy production potential of the dissolved organic content of wastewater would require replacing the commonly used aerobic treatment process with anaerobic treatment. Anaerobic treatment of domestic wastewater has been implemented in a number of developing countries, such as Brazil, Colombia, Mexico, Egypt, and India as a low-cost alternative to aerobic treatment. The limitation for this treatment method is the BOD removal that can be achieved, which could range from 70% to 80%. Typically, the permit requirements require higher levels of removal than can be provided through anaerobic reactors, which results in the need for further treatment before discharge. Anaerobic membrane bioreactors may provide a solution by providing the additional treatment needed within the reactor, offsetting the need for a polishing process. Anaerobic sludge blanket reactors are also an option for anaerobic treatment.

2.3 Kinetic/Hydraulic Content

The term *energy* has different interpretations. In this manual, energy is defined by the science of physics and is the ability or capacity to do work.

There are many different types of energy. These different types can be divided into two broad classifications: potential energy and kinetic energy. Kinetic energy is possessed by a moving object. Potential energy is stored and contained within fuel oil or in an elevated water tank or reservoir. Potential energy is converted to kinetic energy when released. Energy can be changed from one form to another. Common forms of energy are chemical, electrical, mechanical, thermal, radiant, and nuclear. Most people are familiar with the conversion of fossil fuel through burning (chemical energy) to form steam (thermal energy).

Power is the rate at which work is performed. It is the speed at which the work is done or the rate at which energy is expended. The basic unit of power in the International Standard system is the watt, which gives rise to the production of energy at the rate of one joule per second (J/s). In the U.S. customary system, the common units of measure for power are horsepower (1 hp = 550 ft-lb/sec) and British thermal units per hour (Btu/hr). Note that watt and horsepower do not appear to be rate units, but they are. All power units are measured as work per unit time.

The size of a motor or engine is defined not by the total amount of work to be done, but by the rate at which it can do work. Electric motor power is commonly measured in kilowatts (horsepower), boiler output is measured in watts (Btu/hr or boiler horsepower = 33 520 Btu/hr), and air-conditioner power is measured in watts (Btu/hr or tons of refrigeration = 12 000 Btu/hr).

The power multiplied by the time it is applied represents total energy consumption, which is expressed as joules (J) or kilojoules (kJ). Energy consumption is typically measured in joules (kilowatt-hours [kWh]) if consumed as electricity and in kilojoules (British thermal units [Btu]) if consumed as a fuel.

$$\text{Energy (joule)} = \text{power (kW)} \times \text{time (hr)}$$

$$\text{Energy (hp-hr)} = \text{power (hp)} \times \text{time (hr)}$$

$$\text{Energy (Btu)} = \text{power (Btu/hr)} \times \text{time (hr)}$$

$$\text{Energy (ft-lb)} = \text{power (ft-lb/sec)} \times \text{time (sec)}$$

Measured at standard temperature (15.56 °C [60 °F]) and pressure (101.56 kPa [14.73 psi]), natural gas contains approximately 37 000 kJ/m^3 (1000 Btu/cu ft), and billing units are typically 2.8 m^3 (100 cu ft) of gas, 28 m^3 (1000 cu ft or dekatherm), 105 000 kJ (100 000 Btu, otherwise known as therms), or million Btu (MMBTU).

Fuel oil characteristically contains 3.9 × 104 kJ/L (140 000 Btu/gal), and consumption is measured in liters (gallons). Appendix A provides U.S. customary to International Standard conversions for the units used in this manual.

3.0 ELECTRICITY

3.1 Sources

3.1.1 Electric Utilities

3.1.1.1 Electric Service Options

The various service options available will depend on the local utility's circumstance. An example of a nonstandard power purchase arrangement offered by utilities is the interruptible service contract. As discussed previously, utilities sometimes offer lower rates to industrial customers that have the ability to manage their energy consumption.

Interruptible service means that the customer may be required to cease all use of interruptible electrical service when requested to do so by the utility on short notice. Automatic interruptions may be required and are controlled by underfrequency relay equipment.

A typical contract may include provisions that the utility may interrupt power deliveries to the customer no more than "x" times per month or per year, with a specified minimum notification period (e.g., 30 minutes to 1 hour). This contract arrangement provides a discount to the customer in return for giving the utility more flexibility in managing its power deliveries, thus optimizing the utility's use of generation capacity. An onsite power generation facility will be needed at the WRRF.

Curtailable service means that the WRRF may be required to cut back a designated amount of electrical demand (measured in kilowatts) when requested by the utility, on no more than a specified number of occasions.

3.1.2 Onsite Power Supply—Natural Gas

3.1.2.1 Natural Gas Generators

Natural gas generators can be used as emergency energy sources at WRRFs to ensure treatment can continue during power loss events. The benefits of a natural gas generator can include uninterrupted fuel supply, no onsite fuel storage requirements, and less air emissions. Drawbacks include the potential of natural gas line breaks during an emergency situation, which would prevent the operation of the generator.

3.1.2.2 Co-Generation Interconnection and Other Special Agreements

A variety of contracts may be required of customers who operate co-generation facilities.

Examples of special contracts include

- paralleling agreement,
- standby power agreement, and
- standard operating agreement.

A paralleling agreement (also referred to as interconnection agreement) conveys authority to the co-generator to operate its co-generation unit in parallel with the utility power system. *Parallel operation* means operating while connected to the power grid for augmenting power requirements to greater than the generator's abilities to supplement or partially offset the WRRF power demand and consumption. In this contract, the interest of the utility is to protect its network and customers from disturbances that could occur from a co-generator's output. The contract conveys to the co-generator the right to operate in parallel, while allowing the utility the right to place various demands on the co-generator. Some utilities do not favor paralleling because they fear it may have an adverse effect on their grid systems.

A standby power agreement compensates the utility for holding a portion of its generation capacity available to serve WRRF loads should the co-generation unit fail or be taken off-line. A standard operating agreement spells out many of the detailed operating parameters that help maintain the integrity of the utility network. The agreement may also include rules regarding how power plant operations must maintain control and how users should pay for protection equipment that interconnects the two systems.

3.1.3 Onsite Power Supply—Biogas

In the anaerobic digestion process, organic matter is broken down by microorganisms in the absence of oxygen to produce biogas. Biogas is mostly methane (CH_4) CO_2 with small amounts of water vapor and other gases including ammonia and hydrogen sulfide. This is a relatively energy-efficient process. Refer to Chapter 10 for a detailed discussion.

3.1.4 Renewable Energy Sources

Renewable sources of energy include solar photovoltaic, wind, and hydroelectric. Refer to Chapter 11 for a detailed discussion.

3.2 Motor and Drive Basics

3.2.1 Motor Basics

In a conventional building, much of the electricity is used for lighting; however, in a WRRF, most of the electric energy is used in the operation of electrical motors. The increasing cost of electrical power and effect on power grids have driven demand for updating to efficient motors and electrical equipment. Numerous U.S. and international standards organizations have collectively worked to address energy consumption by motors through revision of existing standards or development of new standards. The U.S. Environmental Protection Agency and the U.S. Department of Energy (DOE) have issued many handbooks and guides that promote electrical designs with efficiency as a primary design feature.

The DOE has presented information on the life-cycle cost of a motor, that indicates that the cost of power to operate over a 25-year period may exceed 170 times the initial purchase price of the motor (DOE, 2014). Based on this information and increasing energy costs, the owner and engineer should strongly consider life-cycle cost over initial installation cost. The life cycle might also factor in staging installation of equipment over a period of time to allow for better matching of operating loads. The initial design requirements of motors include maximum expected load, which consider future load levels that may not be reached.

There are numerous types of motors used in a WRRF, from small, fractional, horsepower (hp) motors, to large motors.

3.2.1.1 Components of Common Motors

The design of the electrical motor has a few basic components. Improvements in each component and motor optimization has led to high-efficiency motors.

The armature, or rotor, is part of the motor that is made to rotate by changing the magnetic field. The stator, or field, is the stationary component that causes the magnetic field to change. The shaft, connected to the rotor, provides the primary connection for the conversion of electrical/magnetic energy to rotational force. The air gap is the clearance distance between the rotor and stator. Bearings are designed to minimize the frictional forces of the shaft's rotation. The design and manufacture of each of the electric motor components has to be optimized with the system for energy efficiency.

The system includes mechanical, electrical, and software components. A smart motor package may combine a microprocessor for communications, a controller for loop functionality, imbedded sensors for monitoring motor health, and diagnostic software for operations and performance.

3.2.1.2 Measurements of Electrical Characteristics

The characteristics of the power supplied to a motor and the behavior of the electricity within the motor can be measured. These measurements are used to assess the quality of the power supply, the load being placed on the motor, the motor's general operating efficiency, and the motor's electrical integrity. All electrical tests and measurements must be performed by qualified personnel trained for this technology and requires specialized personal protection equipment.

3.2.1.2.1 Voltage

For diagnostic purposes, three-phase voltage measurements are taken for the three leads (L_1, L_2, and L_3) on the load side of the motor controls. The voltage unbalance for the measurements obtained between L_1 and L_2, L_1 and L_3, and L_2 and L_3 should be less than 1%.

$$\text{Unbalance (\%)} = \text{Maximum deviation average/}\ \text{Average of the three readings} \times 100 \tag{3.1}$$

If the unbalance exceeds 1%, the amperage unbalance and the motor operating temperature will be high. Frequently, problems with the motor controls or power supply result in voltage unbalances. Testing for voltage unbalance should be part of equipment preventive maintenance.

Motors are designed to operate within a narrow voltage range. Typically, the acceptable voltage range for a motor is the nameplate rating ±10%. Voltage tolerance levels are defined by ANSI/IEEE C84.1 (NEMA, 2006a), a joint standard of the American National Standards Institute (ANSI) and the Institute of Electrical and Electronics Engineers (IEEE). The manufacturer's literature for a specific motor should be consulted to ensure proper operation.

3.2.1.2.2 Amperage

The currents (amperage) flowing through each phase (L_1, L_2, and L_3) of a three-phase motor must be close to equal. When the amperage between the phases is unbalanced, the motor runs hot and loses efficiency. Amperage measurements should be taken when the motor is at its normal operating temperature and loading. The percent unbalance is calculated using the same

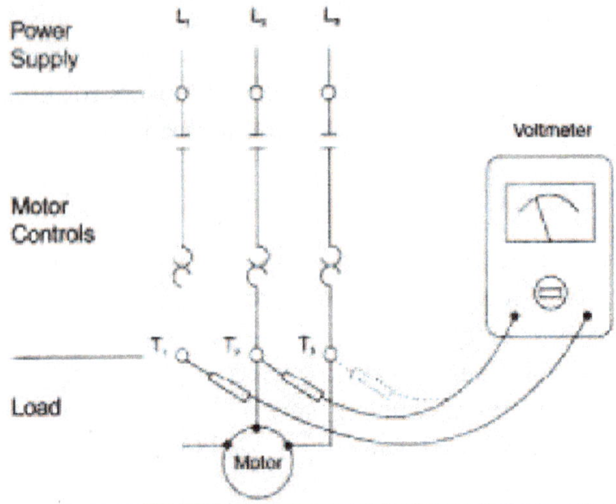

FIGURE 3.1. Checking three-phase motor voltage (L = power lead and T = motor starter terminal). Reprinted with permission from ITT FLYGT Corp. (1985) *FLYGT Product Education Manual,* 8th ed., Product Education, Norwalk, Connecticut.

formula used for voltage unbalance. At full load, the maximum allowable unbalance is 5%. If a motor is operating at considerably less than full load, a larger percentage of unbalance may be acceptable.

Identifying the source of voltage or amperage unbalance requires troubleshooting by trained qualified personnel. First, note the voltage and amperage with the original arrangement of power leads. Next, change the location of all three power leads on the load side of the motor controls and repeat the measurements. It is important to make sure that the positions of all three leads are changed. If only two leads are changed, the motor will operate in the reverse direction. Change all three leads to produce the third possible arrangement and take a third set of readings. If the same lead always produces the high reading, the problem is somewhere between that terminal and the power plant. The problem could be as simple as worn starter contacts or a loose terminal screw. Assistance may be required from the power company for a solution.

3.2.1.2.3 Power Factor

The easiest and most accurate method available to measure the power factor is to use a power factor meter. The power factor is the ratio of the actual power (typically expressed in watts) to the apparent power (typically expressed in voltamperes). This relationship is illustrated by the power triangle shown in Figure 3.2.

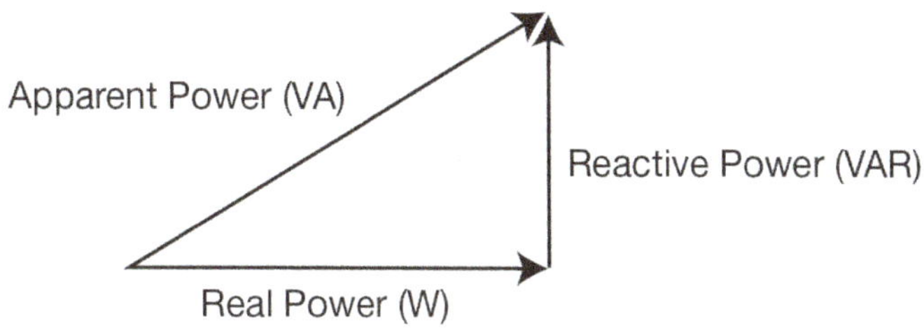

FIGURE 3.2. Motor triangle

The kilovolt-ampere reactive power (kVAR) is the energy used to produce the induced magnetic forces that make the motor rotate. The reactive power is not measurable with a wattmeter. The power factor can also be obtained by calculation, using data obtained with a wattmeter, voltmeter, and ammeter. The real power (kilowatts) and apparent power (kilovolt-amperes) are measured directly and used in the calculation.

3.2.1.2.4 Resistance and Insulation

Note that the tests described in this section need to be performed by a trained professional and require specialized personal protection equipment. The following information is for background only. The flow of electricity from a point of high potential to a point of low potential is governed by the resistance of the space between the two points. An electrical conductor provides a path of low resistance to facilitate the flow of electricity along an intended path with minimal losses. Insulation confines the flow of electricity within its intended path by shielding the electrical path along a conductor with a barrier of high resistance. Over time, the insulation around conductors begins to break down and allows current losses to ground. Insulation on wires leading to the motor and on the motor coils themselves should be checked routinely as part of preventive maintenance. This simple measurement using a megohmmeter ("megger") can help prevent emergency maintenance situations. (Before any testing, it is important to verify that all safety procedures are observed and security is in place.) Because the resistance of the insulation to the flow of electrons is being measured, the resistance values are measured in megohm (million ohm [MΩ]) units. Whenever resistance is being measured, all sources of power must be removed from the circuit being tested. It is important to always check for voltage before putting the megohmmeter probes on the terminals. The megohmmeter has its own power supply to energize the circuit being tested. Each power lead to the motor is checked as shown in Figure 3.3. A new motor should read infinite resistance. Over time, the resistance will gradually decrease. This

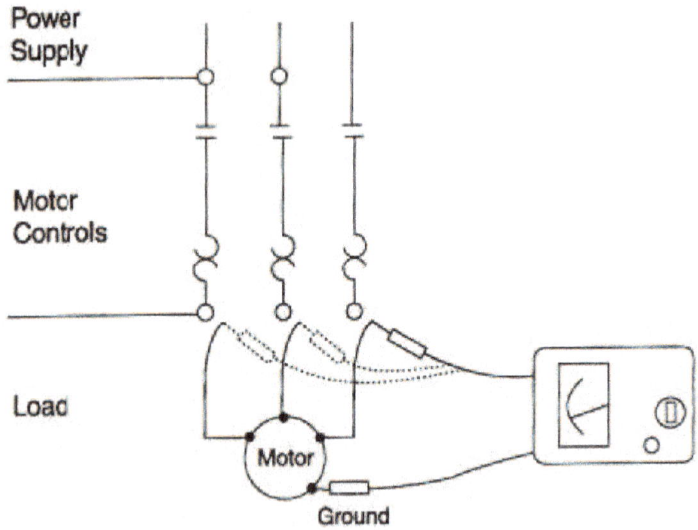

FIGURE 3.3. Checking motor insulation, coil to ground, for a megohmmeter. Reprinted with permission from ITT FLYGT Corp. (1985) *FLYGT Product Education Manual,* 8th ed., Product Education, Norwalk, Connecticut.

information can be graphed so that the trend can be visualized. When a minimum resistance value is reached, the coils can be dipped and rebaked to restore the insulation value. Some manufacturers allow 2 MΩ/hp as the minimum resistance value before the coils are serviced. For small motors, a somewhat higher ratio should be used. If the motor is checked at the motor controls, this activity is also checking the insulation of the conductor to the motor. Before the motor coils are faulted for low resistance, additional readings must be made at the motor. Large motors should consider the use of an online megger.

3.2.1.2.5 Power

Measured power is the actual instantaneous electrical demand required to operate a motor under a full or partial load condition. Power is measured in kilowatts and may be measured directly with a wattmeter. Knowing the power consumed by a motor can be useful in evaluating the operational efficiency of the system. Power measurements can be compared to the motor's power rating to determine whether the motor is overloaded (undersized) or underloaded (oversized). Overloading may be the result of poor motor selection or of a motor, drive, or driven element problem, which requires immediate attention. Motors are most efficient at their full-load condition.

3.2.1.2.6 Slip

Slip is the difference in actual rotor speed and the synchronous speed of the motor, as follows:

$$\text{Slip} = N_S - N \tag{3.2}$$

Where

N_S = synchronous speed, r/min (revolutions/minute); and
N = actual speed, r/min.

Load estimates can be made from motor-operating parameters such as current, slip, and input power. Figure 3.4 shows typical performance curves for a 7.5-kW (10-hp) squirrel cage motor. As can be seen, current is directly proportional to output power in only a narrow range—approximately 70% to 110% of full load. Thus, overestimates of load will be made in cases in which the motor is lightly loaded. Slip, however, is directly proportional to load throughout the operating range and allows a method of load determination at any operating point. Load can also be calculated from input power,

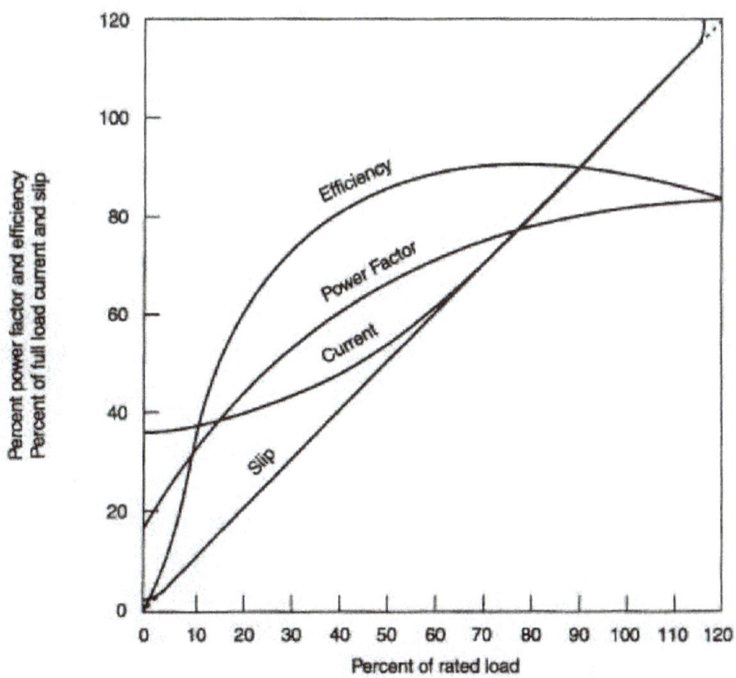

FIGURE 3.4. Typical performance characteristics for a 7.5-kW (10-hp) squirrel cage induction motor as a function of load

but the efficiency at the operating point must be known. Load determination from slip and input power measurements are discussed in the next paragraph. Slip measurements provide a convenient means of estimating motor loading, provided accurate and precise measurement devices are available.

Because slip is proportional to load, the ratio of the measured slip to the full-load nameplate slip yields the relative motor load. The slip of a squirrel cage induction motor is small (typically less than 3%). A stroboscopic tachometer with an accuracy of 0.1% or better is required to produce meaningful results. An error of only 1% would result in an error of 30% in the slip measurement. Furthermore, considerable error may be inherent in this method when using manufacturers' nameplate data because the National Electrical Manufacturers Association (NEMA) standards allow a 20% deviation in the nameplate slip.

3.2.1.3 Operating Power

Motors are rated in units of watts (W) (horsepower) on the basis of the maximum amount of work that they are capable of performing. The actual power consumed at any given time by an operating motor is in direct proportion to the work being performed. In other words, if a 75-kW (100-hp) motor is only doing 60 kW (80 hp) of work, the motor is only supplying the equivalent electrical power of 60 kW (80 hp). In general, motors 10% to 20% larger than actually required are typically selected by the engineer in the design process. It is important to note that the common convention is to use kilowatts in place of watts.

Nameplate power is the output power of the motor; it is also known as the *shaft power*. Input power to the motor is always greater than output power. For example, a 75-kW (100-hp) motor doing 75 kW (100 hp) of work will probably be drawing 80 kW (110 hp) from the power supply. Because motors are commonly oversized by 10% and are only 90% efficient, nameplate power is often taken at face value for power draw in energy calculations.

3.2.1.3.1 Service Factor

The *service factor* is a number placed on each motor by its manufacturer indicating the safe amount greater than the nameplate rating at which the motor can run without failing. The service factor allows for such design operating power requirements as bearing problems. Do not install a motor in an application with a design rating in excess of the motor nameplate. "Service factor" is stamped on each motor nameplate and is expressed as a percentage of the nameplate power rating, for example 1.15 or 115%.

This is the most common service factor. Motor service factors should only be used as a safety factor when the load temporarily exceeds the nominal rating of the motor.

3.2.2 Types of Electric Motors

Motors are available for direct current or alternating current applications, with alternating current motors being available for single-phase or three-phase current loads. Motors are also available for many standard voltages. Small motors are typically single-phase and operate on a 110-V electric current. Larger motors found in WRRFs are typically three-phase and operate on the 480-V three-phase power that normally serves industrial applications. Large motors with higher voltage requirements are needed when greater horsepower is required.

All alternating current electric motors operate on the same basic principles regardless of type or size. Rotation of the shaft is the result of the force created by the interaction of a magnetic field and the current between rotor (rotating electrical coil) and stator (stationary electrical coil). It makes no difference whether the magnetic field is created in the rotor or the stator. Within this simple principle, there are many different types of alternating current motors, each having its own operational characteristics specifically suited to the drive application.

3.2.2.1 Three-Phase Motors

Because of their high efficiency and low maintenance requirements, three-phase squirrel cage induction motors are the most common type in use. The efficiency of squirrel cage motors is affected by a number of factors, including the characteristics of the electrical power supplied and the motor's size, design, electromechanical integrity, and operating load. The last two factors can be partly controlled by the operator through proper operation and maintenance (O&M) practices. Motorized operations should incorporate the most practical energy-efficient systems available during the design process. Comparing competent systems on the basis of annual worth for capital plus O&M costs will facilitate the proper selection.

There are three general types of three-phase motors that all have unique rotors and operating characteristics. The three types are the squirrel cage induction motor, the wound rotor induction motor, and the synchronous motor. The three-phase electric power required for these motors, regardless of whether they are induction or synchronous motors, can be calculated as follows:

$$P = V \times I \times \sqrt{3} \times PF \tag{3.3}$$

Where

P = power, W;
V = line voltage, V;
I = average line current of 3 legs × square root of the number of phases, A; and
PF = power factor.

3.2.2.1.1 Squirrel Cage

The most common industrial motor is the squirrel cage induction motor. It is the least expensive type of induction motor and is available in all common power ratings and synchronous speeds. Voltage is applied directly to the stator or primary winding. The rotor, or secondary winding, consists of bars of aluminum or copper connected together at both ends by a conducting ring in an arrangement resembling a squirrel cage. The rotor windings form a complete closed circuit with no external connections. The absence of moving electrical contacts makes these motors quite reliable.

The speed of a squirrel cage motor is nearly constant over the normal range of loads. The speed of all induction motors is determined by the line frequency, the number of poles in the motor, and the slip. The synchronous speed of the motor is the speed in synchronism with the frequency of the electric current or the speed at which the magnetic field revolves. It depends on the line frequency and the number of poles built into the motor, and is determined as follows:

$$N_s = 120\, f/p \tag{3.4}$$

Where

N_s = synchronous speed, r/min;
f = frequency (60 Hz); and
p = number of poles per phase.

Synchronous speeds are typically available at 900, 1200, 1800, and 3600 r/min. For a squirrel cage motor to produce torque, the operating speed must be less than the synchronous speed. The difference between the operating speed and the synchronous speed, which is called *slip*, is proportional to the torque produced.

Squirrel cage motors are classified by NEMA into four design types (A through D) based on torque, slip, and starting characteristics. Type A has low torque, low slip, and normal starting current. Type B is similar to type A, but draws less current. Type C has more starting torque than types A and B, but cannot bring all loads up to full speed. Type D has high starting torque and high full-load slip.

3.2.2.1.2 Wound Rotor

In the wound rotor induction motor, unlike the squirrel cage motor, the free ends of the secondary windings are brought out to slip rings through brushes and are externally connected. By varying the external resistance, the speed can be controlled. Without external resistance, this motor behaves similarly to the squirrel cage induction motor. Energy is expended in the external resistance, and efficiency varies approximately in proportion to the speed reduction. Use of the wound rotor motor is typically reserved for applications in which high starting torques and intermittent operations are required, such as for hoists and cranes.

3.2.2.1.3 Synchronous Motors

In a synchronous motor, direct current is applied to the rotor, which locks in step with the rotating magnetic field of the stator causing the rotor to rotate at synchronous speed. However, the synchronous motor, by itself, is incapable of producing torque at other than synchronous speed; thus, alternative starting mechanisms, such as a squirrel cage winding, are required. In the industrial field, synchronous motors are used only in the larger power range, typically 75 kW (100 hp) or greater.

In contrast to the induction motor, a synchronous motor's power factor is not a function of load or size but is under the control of the user. By adjusting the direct current, the power factor can be changed to a leading, lagging, or unity value. Thus, synchronous motors may be used to offset the power factors of inductive loads. The permanent magnet, brushless, synchronous motor has become an industry standard in blower applications. They are also starting to show up in pump applications.

3.2.2.2 Single-Phase Induction Motors

Single-phase current is not typically produced in the United States, but it is readily derived from any two lines of a three-phase supply system. Because of the higher costs associated with local transmission and voltage step-down of three-phase current, single-phase, low-voltage current is supplied to homes and farms. Unlike three-phase motors, single-phase motors are incapable of self-starting and require an auxiliary winding. Motors for single-phase current are more complex than three-phase motors because they require additional devices such as switches and capacitors to induce the rotating magnetic field. Consequently, single-phase motors are generally more expensive, less efficient, less reliable, and have shorter service lives than three-phase motors. For many small-power applications, however, the extra motor cost is outweighed by the costs of additional wiring and motor controls. Additionally, low-power, single-phase motors feature the added convenience of being able to be plugged into the nearest outlet. There are many types of single-phase motors, including shaded-coil, inductive-split-phase, capacitor, repulsion-start/induction-run, and repulsion-induction. For industrial applications, capacitor motors are typically used. The most common capacitor motors are the capacitor-start motor, the capacitor-start/capacitor-run motor, and the permanent-split-capacitor motor.

Although the single-phase motor may not be the best fit for most applications, it has been found that some smaller remote sites may only have single-phase motors available. Using single-phase motors will typically result in higher energy costs and possibly shorter motor life; however, in these special cases there may be no option. Designers should strive to use the highest quality motors in these situations to minimize maintenance issues. Small fraction horsepower applications may also be a good fit for single-phase motors where availability is a factor. These motors have low current draw and have minimum effect on overall energy usage. An alternate approach to the use of three-phase motors with single-phase power availability is to use a phase converter and/or variable frequency drive. Although not as efficient, this approach will provide a mechanism for a remote application.

3.2.2.3 Direct Current Motors

An advantage of direct current motors is precision control. Because of this, they are often used as controllers in automation applications. They are also used for other applications requiring precision, such as cranes, hoists, and elevators. Additionally, they are used in battery-powered mobile equipment and for many electric railway operations. In WRRFs, large direct current motors have been used in variable-speed control applications such as influent and return activated sludge pumping systems using rheostats or variable-voltage controllers. Because it is impractical for large power producers to generate and distribute direct current voltage, voltage for direct current motors is typically derived near the application from alternating current through various types of rectifiers.

3.2.3 Motor Efficiency and Performance

3.2.3.1 Definition of Efficiency

Motor efficiency is the ratio of motor output to motor input. It is a measure of how well a motor converts electrical energy to mechanical energy and is typically expressed as a percentage.

$$\text{Motor efficiency} = P_m/P_e \times 100 \tag{3.5}$$

and

$$P_m = P_e - P_L \tag{3.6}$$

Where

P_m = mechanical power output, W;
P_e = electrical power input, W; and
P_L = power loss, W.

Power loss is that part of the input power that is converted to heat rather than useful work. This loss can be attributed to friction and windage, electrical resistance losses in the rotor and stator, magnetic core losses (hysteresis and eddy currents), and stray load losses.

3.2.3.2 Test Procedures

The IEEE standard IEEE-112 (IEEE, 2017) is a recognized test procedure in the United States that is used to determine motor efficiency. Caution should be exercised when comparing efficiencies based on U.S. standards with those of foreign manufacturers as there are differences in international testing methods that result in different levels of premium motors.

3.2.3.3 Matching Motors to Load

The primary goal in selecting a motor is to be sure the motor selected will have adequate power to drive the intended load throughout the entire anticipated load range. A secondary goal is that the motor be energy efficient and economical.

Motor efficiency for standard induction-type motors increases with increasing motor size and is relatively constant for a given motor at various loads. However, most motors are slightly more efficient at full load, with efficiency dropping off slightly as load decreases from full to half load. For motors of greater than 0.75 kW (1 hp), the change in efficiency from full to half load is generally less than 5% of the full-load efficiency. Similarly, a power factor for a standard induction motor increases with motor size, but the power factor drops off more significantly with a reduction in load. The decrease in power factor for a 0.75-kW (1-hp) motor at half load is approximately 25% of the power factor at full load.

Power factors decrease 10% to 15% or more as load is reduced from 100% to 50%, depending on the power and revolutions per minute of a given motor. Because electrical utilities often assess charges for a poor power factor, it is advantageous to closely match motor size to load to maintain a high power factor at a WRRF. The relationship of both efficiency and power factor to changes in motor load differ between the motor enclosure (open drip-proof, totally enclosed fan-cooled, and others), number of revolutions per minute, and manufacturer.

When studying equipment, if a significant mismatch in the load and electric motor rating is found, it may be economical to replace the motor in question with a motor that more closely approximates the required load.

3.2.3.4 Standards for Energy-Efficient Motors

As defined previously, *motor efficiency* is the ratio of mechanical power output to the electrical power input, typically expressed as a percentage. Numerous U.S. and international standards now define different versions of high-efficiency motors. This section will focus on the primary standards used in the United States.

High-efficiency motors differ from standard-duty motors in terms of higher quality materials, more precise machining and manufacturing, more precise control over air gaps and tolerances, better bearings, more copper in windings, and extended length. This premium cost typically allows for better warranty protection from most manufacturers. Because of these improvements, energy-efficient motors have higher service factors, longer insulation and bearing life, lower waste heat output, and less vibration.

The following section cites agencies that are currently involved with energy efficiency for electric motors and their key programs. However, energy efficiency is a rapidly developing area of interest and new information is continually being added. As such, the reader should check with the agencies for the most current information.

3.2.3.4.1 Energy Policy Act of 1992

Because motor operation constitutes a substantial percentage of overall energy consumption, Congress granted DOE the authority to set minimum efficiency standards for certain classes of electric motors. Congress enacted the Energy Policy Act of 1992 (EPAct), which focused on reducing the total amount of energy used in the United States. The Energy Policy Act's rules for motors became effective on October 24, 1997. However, these rules did not immediately bring about new motor designs. The Energy Policy Act also granted large exclusions for existing motors.

3.2.3.4.2 Independent Energy Companies

There are several independent companies that have been established as nonprofit corporations that focus on energy efficiency. One such company,

Advanced Energy (Raleigh, North Carolina), is an example of an independent nonprofit corporation that is governed by a board of directors appointed by the governor of North Carolina and member utilities. Numerous motor-related programs are available through Advanced Energy.

3.2.3.4.3 International Motor Standards

As in the United States, the international market has also addressed energy efficiency. This has resulted in many new international standards as well as revisions of old standards. However, it is beyond the scope of this chapter to provide a reasonable review of all the current international motor standards that affect design efficiency. It should be noted that several of these standards have different methods of determining their "premium" motor that are different from the "NEMA Premium" motor. As such, the engineer should closely examine these differences to determine if they still meet their design criteria.

3.2.3.4.4 National Electrical Manufacturers Association

Although the minimum standards set by DOE were reasonable at the time they were drafted, industry noted the need for more efficient motors. In 2001, NEMA established standard MG-1, which defined energy-efficient "premium" electric motors; the standard was later updated in 2006 (NEMA and ANSI, 2006b). These standards were more specific and more stringent than those required by EPAct motors. The NEMA Premium efficiency electric motor program defines the parameters products must meet or exceed to be considered a NEMA Premium electric motor. The NEMA Premium efficiency levels are contained in Tables 12–12 and 12–13 in NEMA Standards Publication MG-1-2006.

To be considered energy efficient, the motor's performance must equal or exceed the nominal full-load efficiency values provided by this standard. Specific full-load nominal efficiency values are provided for each horsepower, enclosure type, and speed combination. The motor's performance must equal or exceed the efficiency levels given in this standard.

The Energy Policy Act of 1992 requires that most general purpose motors manufactured for sale in the United States now meet this standard. This specifically covers motors up to 1200 hp (900 kW) general purpose, T-Frame, single-speed, foot-mounted, continuous-rated, polyphase, squirrel cage induction motors conforming to NEMA Design A and B. Covered motors are designed to operate with 230 or 460 V power supplies, have open or closed enclosures, and operate at speeds of 1200, 1800, or 3600 rpm (revolutions per minute).

TABLE 3.1. Nominal efficiencies for "NEMA Premium®" induction motors rated medium volts 5 kV or less (form wound). Tables reprinted by permission of the National Electrical Manufacturers Association, Rosslyn, Virginia. NEMA MG1-2006. NEMA Premium®.

	OPEN MOTORS					
	2 POLE		4 POLE		6 POLE	
HP	Nominal Efficiency	Minimum Efficiency	Nominal Efficiency	Minimum Efficiency	Nominal Efficiency	Minimum Efficiency
250	94.5	93.6	95.0	94.1	95.0	94.1
300	94.5	93.6	95.0	94.1	95.0	94.1
350	94.5	93.6	95.0	94.1	95.0	94.1
400	94.5	93.6	95.0	94.1	95.0	94.1
450	94.5	93.6	95.0	94.1	95.0	94.1
500	94.5	93.6	95.0	94.1	95.0	94.1
	ENCLOSED MOTORS					
	2 POLE		4 POLE		6 POLE	
HP	Nominal Efficiency	Minimum Efficiency	Nominal Efficiency	Minimum Efficiency	Nominal Efficiency	Minimum Efficiency
250	95.0	94.1	95.0	94.1	95.0	94.1
300	95.0	94.1	95.0	94.1	95.0	94.1
350	95.0	94.1	95.0	94.1	95.0	94.1
400	95.0	94.1	95.0	94.1	95.0	94.1
450	95.0	94.1	95.0	94.1	95.0	94.1
500	95.0	94.1	95.0	94.1	95.0	94.1

3.2.3.4.5 Institute of Electrical and Electronic Engineers

The NEMA standard MG-1 recognizes the motor efficiency testing protocol established by IEEE 112 (IEEE, 2004) as the basis of testing for energy-efficient motors. The IEEE 112 standard test procedure recognizes five methods for determining motor efficiency. Each method bears certain advantages over the others with respect to accuracy, cost, and simplicity. The NEMA MG-1 references IEEE 112 method B. This test uses a dynamometer to measure motor output under load. Different testing methods yielding slightly different results are used in other countries. The NEMA nameplate labeling system for design of A and B motors in the 10 to 500 hp (7 to 370 kW) range uses bands of efficiency values based on IEEE 112 testing.

3.2.3.4.6 Consortium for Energy Efficiency

Consortium for Energy Efficiency is a nonprofit organization that includes many electrical utilities among its members. This group recognizes NEMA Premium motors up to 200 hp (150 kW) as meeting their criteria for possible energy-efficiency rebates.

The lack of labeling indicating a NEMA Premium motor does not necessarily mean the motor does not meet a high-efficiency standard. Some companies are not members of NEMA and, therefore, cannot apply the label. The owner may use other means to determine motor efficiency. Federal agencies are required to purchase NEMA Premium motors where applicable.

3.2.4 Motor Management

Motor management should include proper operation and maintenance to remain energy efficient and in optimal operation. Motor management may include but is not limited to maintaining up-to-date motor records, specifications, inventory tracking and maintenance of power quality, ensuring proper lubrication of the motors, and proper protection of the motor from environmental damage. Frequency checks of the mechanical operation and drive equipment for vibration, noise, high temperature, or other abnormal conditions and providing proper routine maintenance, dust removal, and ensuring proper lubrication are extremely important for guaranteeing a long motor life.

The electrical operation of the motor affects the cost of operation and the ability of the motor to provide dependable service. The motor should be checked for proper voltage, amperage, and resistance to ground. Operation of the motor controls should also be checked carefully. Motor starters should engage fully and not chatter. Weak coils and worn contacts should be replaced. Thermal sensors and cameras can be used to check for the proper dissipation of heat on the contacts and to verify the connection tightness. Once per year, all connections in the control panel should be checked for tightness.

The motor repairs must be addressed. The exact cause of failure should be determined by qualified personnel. Before ordering a direct replacement from the manufacturer, the life cycle and energy efficiency of the motor should be considered. It may be found that the motor should be upgraded using guidelines set for energy efficiency.

All motor repair facilities are not equal. Bringing a damaged motor back to its "as-built" condition requires a level of skill that is not available in all repair shops. A poorly repaired motor can easily be a less efficient motor, thus defeating the facility's energy efficiency program. Before committing to a specific repair, an audit of the motor repair facility should be conducted as follows:

- Determine the condition of the equipment that will be used for the repairs.
- Determine if proper safety standards for shop operation are in effect.
- Determine if proper record keeping is maintained.
- Determine the knowledge level of the staff involved with the repair.
- Determine that the repair facility stocks and uses the proper wire and components that are needed.
- Check the test equipment.
- Check if the repair facility complies with motor repair quality standards or equipment specifications.
- Check if the facility complies with standards and procedures published by the Electrical Apparatus Service Association.
- Determine if the facility has a written quality program.
- Check to see if the facility has a lubrication program, as motor efficiency and bearing life both suffer from over-greasing.

3.2.5 Variable Speed Drives

Based on application and load characteristics, the motor may need a drive, which is a form of pump speed control. Drives and control systems are selected for greater equipment use. The benefit of a good operating drive allows for greater motor reliability, reduced equipment wear and tear, decreased maintenance cost, and increased capacity utilization. Two common variable speed drives found in a wastewater application are the variable frequency drives and eddy current drives, discussed below.

3.2.5.1 Variable Frequency Drive

Variable frequency drives are used for controlling the rotational speed of an alternating current electric motor by controlling the frequency and voltage of the electrical power supplied to the motor. Variable frequency drives are used to control both induction and synchronous motors. Based on the system inputs, such as flow and pressure, the control system allows for the setting of motor speed and the option for start/stop functions.

3.2.5.2 Eddy Current Drive

The eddy current drives consist of a fixed speed motor and an eddy current clutch. The clutch contains a fixed speed rotor and an adjustable speed rotor separated by a small air gap. A direct current in a field coil produces a magnetic field. Varying the clutch current allows the clutch to transmit enough torque to operate at the desired speed. The variation allows for the controller to provide closed loop speed regulation. Speed feedback is provided by an integral alternating current tachometer.

4.0 NATURAL GAS

4.1 Uses for Natural Gas in a Water Resource Recovery Facility

Natural gas is typically used in boilers and unit heaters as a heating source at a WRRF. As mentioned in the previous section, natural gas can also be used to power a generator to create electricity.

4.2 Sources

Like the electrical service options, the various service options available for natural gas will depend on the local utility's circumstance. Natural gas utility service is typically limited to locations where natural gas infrastructure is available to be connected to.

4.3 Utility Bills

The basic rate components of demand, energy, and customer charge are similar in concept for natural gas with that of other sources of energy. In addition, the same variations on rates, based on peak day and season, are typically applied.

With natural gas, energy is measured by units of gas consumed, measurable in hundreds of cubic feet (ccf) (cu ft $\times$ 0.02832 = m^3), thermal heating value (therms), or millions of British thermal units (MMBtu) (1.055 GJ):

- 1 therm = 100 000 Btu,
- 1 MBtu = 10 therms, and
- 1000 ccf (MCF) equals approximately 1 000 000 Btu or 1 dekatherm.

Because 0.03 m^3 (1 cu ft) of methane has a heating value of approximately 1055 kJ (1000 Btu), 3 m^3 (1 ccf) of natural gas has a heating value of approximately 1 therm.

Recently, the trend has been to offer separate rates for the commodity itself (therms of natural gas) and its associated pipeline transportation charges and for local distribution of the commodity through the local utility's pipelines to the customer. These changes in rate structure are in response to deregulation and to a growing number of independent gas producers that offer natural gas with a variety of pricing structures, but whose existing distribution lines are owned and operated by established utilities to deliver the commodity to the customer. In addition, there could be other charges such as balancing, storage fuel charges (for operating compressors), and losses in the pipeline transportation from the hub.

4.3.1 Rate Structures

Billing determinants for gas production and supply services vary from state to state and also depend on the industry arrangement present in the local market. The following is a general view of the diversity of industry status.

4.3.1.1 Unbundled Utility Service

Through deregulation, competitive markets have developed in some states as alternatives to the traditional monopolistic markets of local gas suppliers. A "noncore" distinction is made for customers that have alternative choice(s) for procuring gas or other economic fuel alternatives. In the noncore market, gas services have been disaggregated so that gas procurement, gas transportation, delivery balancing, and storage services can be separately procured and charged.

4.3.1.2 Transportation

Some states distinguish gas services between traditional procurement-transportation-supply service and transportation service only.

4.3.1.3 Pipeline Direct

A third condition exists in some areas where the Federal Energy Regulatory Commission use regulated interstate pipelines to service industrial demands with a direct connection from a transmission line and bypass the local distribution company serving the surrounding area.

The billing components of gas and transportation services are similar to those of electric utilities as discussed previously:

- *Demand* is the contracted level of service indicated by the customer's experienced peak usage on the system's peak day or a negotiated maximum until operating experience justifies a change. Demand may also be assessed using monthly peak or moving average peaks to assess the system service rendered.
- *Demand charge* is a monthly fixed charge that is applied to the measured or aforementioned contract demand.

- *Energy*, or *throughput*, is a variable rate assessed on the actual gas throughput that has been conveyed or delivered.
- *Interruptible service* in the gas industry is typically limited to large customers. Interruptions are often based on community heating load conditions. The level of interruption is a negotiable element in service rates.

4.3.2 Rates

Rate categories vary for each utility with regard to terminology and the conditions that apply to each class of service. General service rates apply to small users, are typically at a firm rate, and are not curtailable in many parts of the country. Industrial rates vary more seasonally, may require a minimum usage per month through the year, and vary in curtailability with the utility. Large users will find industrial rates attractive for cost savings, particularly in summer months, but the frequency of curtailability must be made viable by either reducing usage or switching to a backup fuel. Additionally, unbundling of natural gas rates in some states has resulted in the availability of special contracts for natural gas transportation and supply.

4.3.2.1 Gas Transportation Agreement

The gas transportation agreement covers transportation services the utility provides in moving gas from the state border (or in-state gas fields) to the customer delivery point. The contract may provide that the customer will be served under a co-generation transportation rate schedule, which provides that the transport cost must be no more than the equivalent service to the utility's gas boilers that generate electricity. The rate structure has components that vary with the volume of gas used and measured monthly and other components that fluctuate on a rolling basis of 12 months, measured on the highest demand at any single point in time.

Gas companies typically offer transportation programs that may be attractive to large year-round users. These are similar to commodity contracts in that the user agrees to buy a specific quantity in a specific time period at a specific price. Cost savings for a large user can be significant; however, the risk of committing to price, quantity, and delivery time must be carefully weighed.

The transportation of gas from well field to the local distribution company can be performed by the purchaser (in the case of large industrial loads) or by a gas broker. Once the transportation route and user's operating needs are addressed, the local gas company will assist with ultimate delivery.

The local gas company agrees to arrange transportation from the pipeline interconnect to the user typically for a fixed fee per million British thermal unit (MBtu) and a fixed monthly administrative charge. The gas company may require that the customer commit to a specific quantity of gas to be delivered for the following month by day 5 of each month. Gas purchased under this type of program is typically curtailable to 0%. In addition, there will typically be a minimum quantity per month to the program, and gas quantities used beyond the contract amount will be charged an industrial rate.

To illustrate how supply and transportation of natural gas can be billed, Table 3.2 presents an example of a separate natural gas billing for the supply of natural gas to the city gate, and Table 3.3 presents an example of a separate natural gas billing for the transportation from the city gate to an Upper Midwest WRRF in 2008.

4.3.3 Seasonal Pricing Incentive

Prices for natural gas vary with the market conditions for supply and demand. Because demand is high in winter, prices are typically highest then and lowest in the summer months. Unfortunately, WRRFs fall into this pattern of high demand in winter because space heating needs and process demands such as sludge heating peak in colder months. Some WRRFs may

TABLE 3.2. Example of natural gas billing for supply

Charge Item	Use	Units	Rate ($)	Charge
Supply	15 886.45	Dekatherm	9.44	$149,968.09
Balancing and Nominating	15 886.45	Dekatherm	0.05	$794.32
Total				$150,762.41

TABLE 3.3. Example of natural gas billing for transportation

Charge Item	Use	Units	Rate ($)	Charge
Customer Charge	Large			$673.17
Distribution Margin	158 864.50	Therms	0.0343	$5,449.05
Demand	8110.00	Therms	0.1475	$1,196.23
Total				$7,318.45

Note: In this example, the total average cost of supply and transportation of natural gas was $0.995 per therm.

have peak electrical power requirements in summer and may try to take advantage of low summer natural gas prices (or digester gas produced on site) to fuel engines that drive generators, pumps, or blowers. If this method is used with natural gas to *peak shave* or provide *prime* power, a more constant pattern of usage may be established over the entire year to take advantage of off-peak rates available from gas companies. A WRRF with an anaerobic digestion system that produces digester gas may need natural gas for only the coldest months; thus, it will be difficult to get anything but natural gas at the highest rates for the limited time of usage. In any case, it would be a good idea to look at the WRRF's gas usage patterns and discuss them with the gas company representative at least every 5 years to make sure that the WRRF is getting the best rate available to meets its needs.

4.3.4 Computing Thermal Consumption

Calculating the therms used requires a knowledge of the volume of gas used as well as the heating value. For natural gas, the heating value is available from the gas company and will normally range from 37 000 to 40 000 kJ/m^3 (1000 to 1080 Btu/cu ft). Liquid petroleum gas heating values are also available through the gas supplier and are typically 26 $\times$ 10^6 kJ/m^3 (92 000 Btu/gal).

Because gas is a compressible fluid, it is not always sufficient to know the volume of gas without knowing its pressure. Natural gas is supplied at varying pressures, and the compressed volume delivered must be converted to a volume at standard conditions of temperature and pressure. It should also be noted that the moisture content and hydrogen sulfide concentration are important in determining the corrosiveness of the gas.

For natural gas customers with an interruptible service, a backup fuel is required. For equipment such as burners for boilers, a dual fuel burner will serve this purpose. For smaller equipment such as space heaters, different burner orifices are required to fire liquid petroleum gas and different burners (and heaters) for fuel oil. Therefore, it is important to have a portion of the non-interruptible service to serve those smaller units that cannot be operated with dual fuels.

5.0 OTHER SOURCES OF ENERGY

5.1 Diesel

Diesel will be used in small quantities to fuel emergency standby generators. It may also be used to fuel vehicles.

5.2 Propane

Propane may be used on site for heating, in lieu of natural gas.

5.3 Steam

Some WRRFs make their own steam or purchase as a utility from an "over the fence" source. Steam may be used in solids conditioning, to heat the solids; in dryers to dry the solids. It may also be generated by heat recovery and then used as a motive force for steam powered equipment. It may also be used for building heat.

6.0 REFERENCES

Institute of Electrical and Electronics Engineers. (2004). IEEE 112 Motor efficiency testing protocol.

Institute of Electrical and Electronics Engineers. (2017). Institute of Electrical and Electronics Engineers standard IEEE-112.

National Electrical Manufacturers Association. (2006a). ANSI/IEEE C84.1.

National Electrical Manufacturers Association. (2006b). NEMA ANSI C84.1:2006.

U.S. Department of Energy. (2014). *Premium efficiency motor selection and application guide, a handbook for industry*. U.S. Department of Energy, Energy Efficiency & Renewable Energy, Advanced Manufacturing Office, DOE/GO-102014-4107, February 2014, p. 4-2.

7.0 SUGGESTED READINGS

The following publications were prepared by independent sources. Neither the Water Environment Federation, their volunteers, contractors, subcontractors, or employees makes any warranty, express or implied, or assumes any legal responsibility for the accuracy, completeness, or usefulness of any information, apparatus, product, or process described in the following references. In addition, no product endorsement is implied by the use of examples, figures, or courtesy photos.

Commission of the European Communities. (2009). Commission Regulation (EC) No. 640/2009, of 22 July 2009, implementing Directive 2005/32/EC of the European Parliament and of the Council with regard to ecodesign requirements for electric motors. https://eur-lex.europa.eu/legal-content/EN/TXT/?uri=CELEX%3A32009R0640 https://ec.europa.eu/growth/single-market/european-standards/harmonised-standards/ecodesign/electric_motors_en

Commission of the European Communities. (2019). Commission Regulation (EU) 2019/1781, of 1 October 2019, laying down ecodesign requirements for electric motors and variable speed drives pursuant to Directive 2009/125/EC of the European Parliament and of the Council, amending Regulation (EC) No 641/2009 with regard to ecodesign requirements for glandless standalone circulators and glandless circulators integrated in products and repealing Commission Regulation (EC) No 640/2009. https://eur-lex.europa.eu/legal-content/EN/TXT/HTML/?uri=CELEX:32019R1781&from=EN

The Electro-Mechanical Authority. (2003). *The effect of repair/rewinding on motor efficiency*. EASA/AEMT Rewind Study and Good Practice Guide to Maintain Motor Efficiency, December 2003. https://easa.com/Resources/Resource-library/the-effect-of-repairrewinding-on-motor-efficiency

U.S. Department of Energy. (n.d.). *Determining electric motor load and efficiency*. U.S. Department of Energy, Motor Challenge, Fact Sheet, DOE/GO-10097-517. https://www.energy.gov/sites/prod/files/2014/04/f15/10097517.pdf

U.S. Department of Energy. (n.d.). *Replacing an oversized and underloaded electric motor*. U.S. Department of Energy, Motor Challenge, Fact Sheet, DOE/GO-10096-287. https://www.energy.gov/sites/prod/files/2014/04/f15/mc-2463.pdf

U.S. Department of Energy. (2000). *Motor repair tech brief*. Office of Industrial Technologies, Office of Industrial Technologies, Energy Efficiency and Renewable Energy, U.S. Department of Energy, DOE/GO-10099-938, March 2000. https://www.energy.gov/sites/prod/files/2014/04/f15/tech_brief_motors.pdf

U.S. Department of Energy. (2012a). *Adjustable speed drive part-load efficiency*. U.S. Department of Energy, Energy Efficiency & Renewable Energy, Advanced Manufacturing Office, Energy Tips: Motor Systems, Motor System Tip Sheet #11, DOE/GO-102012-3730, November 2012. https://www.energy.gov/sites/prod/files/2014/04/f15/motor_tip_sheet11.pdf

U.S. Department of Energy. (2012b). *Eliminate voltage unbalance*. U.S. Department of Energy, Energy Efficiency & Renewable Energy, Advanced Manufacturing Office, Energy Tips: Motor Systems, Motor System Tip Sheet #7, DOE/GO-102012-3733, November 2012.

U.S. Department of Energy. (2012c). *Estimating motor efficiency in the field.* U.S. Department of Energy, Energy Efficiency & Renewable Energy, Advanced Manufacturing Office, Energy Tips: Motor Systems, Motor System Tip Sheet #2, DOE/GO-102012-3734, November 2012. https://www.energy.gov/sites/prod/files/2014/04/f15/estimate_motor_efficiency_motor_systemts2.pdf

U.S. Department of Energy. (2012d). *Is it cost-effective to replace old eddy-current drives?* U.S. Department of Energy, Energy Efficiency & Renewable Energy, Advanced Manufacturing Office, Energy Tips: Motor Systems, Motor System Tip Sheet #12, DOE/GO-102012-3737, November 2012.

U.S. Department of Energy. (2012e). *Magnetically coupled adjustable speed motor drives.* U.S. Department of Energy, Energy Efficiency & Renewable Energy, Advanced Manufacturing Office, Energy Tips: Motor Systems, Motor System Tip Sheet #13, DOE/GO-102012-3738, November 2012.

U.S. Department of Energy. (2012f). *When to purchase premium efficiency motors.* U.S. Department of Energy, Energy Efficiency & Renewable Energy, Advanced Manufacturing Office, Energy Tips: Motor Systems, Motor System Tip Sheet #1, DOE/GO-102012-3743, November 2012. https://www.energy.gov/sites/prod/files/2014/04/f15/whentopurchase_nema_motor_systemts1.pdf

U.S. Department of Energy. (2014a). *Improving motor and drive system performance, a sourcebook for industry.* U.S. Department of Energy, Energy Efficiency & Renewable Energy, Advanced Manufacturing Office, Advanced Manufacturing Office, DOE/GO-10014-4356, February 2014.

U.S. Department of Energy. (2014b). Premium efficiency motor selection and application guide, a handbook for industry. U.S. Department of Energy, Energy Efficiency & Renewable Energy, Advanced Manufacturing Office, DOE/GO-102014-4107, February 2014. https://www.energy.gov/sites/prod/files/2014/04/f15/amo_motors_handbook_web.pdf

4

An Overview of Energy Use at Water Resource Recovery Facilities

Andy Shaw, PhD, PE; Arifur Rahman, PhD, PE; Peter V. Cavagnaro, PE, BCEE, PMP; & Thomas E. Jenkins, PE

1.0 EFFECT OF WASTEWATER CHARACTERISTICS AND DISCHARGE PERMIT REQUIREMENTS

1.1 Introduction

The overall energy use at a water resource recovery facility (WRRF) is a balance between what is used on site to treat the wastewater and what can be generated, or recovered, from the wastewater itself. A "stronger" wastewater may take more energy to treat, but also contains higher concentrations of constituents that can be recovered to generate the energy needed for treatment. Understanding the makeup or "characteristics" of the wastewater is fundamental to understanding the energy balance.

1.2 Wastewater Characteristics

1.2.1 Carbon Content

The carbon content of wastewater is typically measured using biochemical oxygen demand (BOD) or chemical oxygen demand (COD), although sometimes the measurement of total organic carbon (TOC) or total carbon can also be used. Whatever measurement is used, a higher concentration means a higher potential for harnessing that carbon for energy. The ability to capture and recover the carbon energy depends on the form it takes. In general terms, Figure 4.1 shows a typical breakdown for the carbon in regards to its solubility and biodegradability.

If the carbon compounds are not captured and harnessed for recovery, energy is required to treat and convert them to biomass or carbon dioxide. In general, particulate carbon can be captured and transferred to sludge and directed to biosolids processing. Soluble biological carbon is most commonly treated using energy-intensive aerobic treatment, though it can be treated using anaerobic treatment alternatives. The "colloidal" carbon is material that does not typically settle and may not be easily filtered but can be coagulated to form solids that can be removed. The conversion of carbon to a form that can be used as a source of energy is described in Chapter 10.

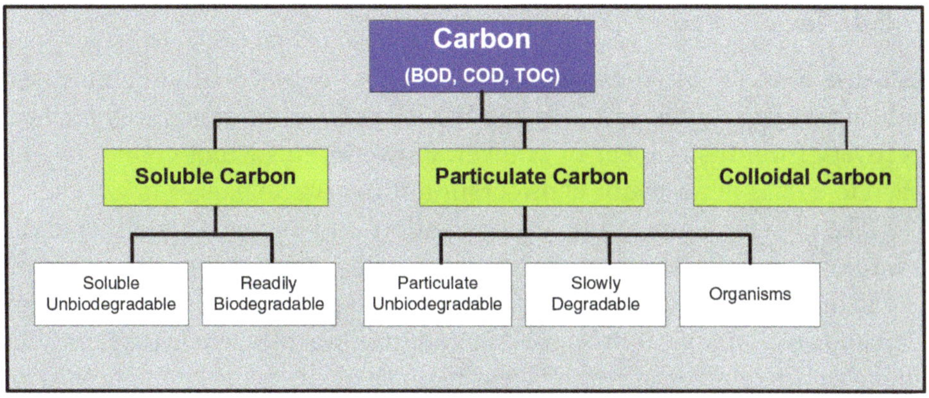

FIGURE 4.1. Figure showing carbon types based on solubility and biodegradability

1.2.2 Nitrogen

Ammonia-nitrogen conversion to nitrates is commonly achieved using biological nitrification, which is an energy-intensive aerobic process. A WRRF that has higher concentrations of ammonia or organic nitrogen (which is converted to ammonia in biological treatment) in the influent and low discharge limits will typically have a higher energy demand than a corresponding facility. Nitrogen removal typically uses biological denitrification, which, in most cases, reduces energy requirements for treatment using nitrates instead of oxygen to remove organics. It is advantageous to incorporate denitrification into a treatment process for energy recovery and other benefits, even if nitrogen removal is not required in a simple nitrifying process.

Total nitrogen removal requires carbon and, therefore, any facility looking to recover energy must pay careful attention to the ratio of carbon to nitrogen in the influent. In some instances, additional carbon is required for denitrification.

1.2.3 Phosphorus

Phosphorus removal, either through biological or chemical means, does not typically have a large energy effect, and, therefore, the concentration of phosphorus in the influent does not have a direct effect on energy.

If removed by chemical means, there is a cost of chemicals and solids processing, and, therefore, the removal by biological means may be preferred. Biological phosphorus removal requires carbon. If the facility wants to recover energy, attention must be given to the ratio of carbon to phosphorus in the influent. In some instances, additional carbon must be added if the ratio is too low.

1.2.4 Temperature

Because of the thermal capacity of water, heat is potentially the most significant energy source and sink in a WRRF and can be considered for heat recovery from the influent or effluent. However, this is not common, and, therefore, it is not a major consideration for energy balances in general.

The influent temperature has a big effect on biological activity and the rate at which biological conversions occur, with warmer temperatures resulting in higher activity, higher reaction rates, and potentially increased energy demands for biological processes. In addition, warmer water temperatures make aerated processes less efficient because of the decreased solubility of oxygen. Many solids treatment processes require heat (e.g., anaerobic digestion, incineration). Processes within the facility may look to recover heat, particularly in solids processing.

Some facilities do have a permitted maximum temperature for effluent discharge and in some cases require cooling for their effluent (e.g., using cooling ponds), particularly industrial treatment facilities with warmer wastewater process streams.

1.3 Discharge Permit Requirements

In addition to considering the inputs to the WRRF, the energy use is also dependent upon the extent of treatment required, some of which has been touched on in the discussion on influent wastewater characteristics.

Water resource recovery facilities that provide secondary treatment alone may have discharge limits of 30 mg/L for BOD and total suspended solids (TSS) removal and typically have the lowest energy footprint overall. Facilities that use low energy processes such as trickling filters have particularly low energy footprints. With anaerobic digestion and strong enough wastewater, these facilities have the potential to be energy neutral overall (WRF, 2015).

Facilities with an effluent ammonia limit require nitrification, which is a significant increase in aeration energy, typically doubling the mass flow of oxygen required for treatment in an activated sludge process. The addition of an anoxic zone (with internal recycle) can offset this effect and recover alkalinity, making this a good approach for facilities even if they do not have a nitrogen or nitrate limit.

As noted above, phosphorus removal does not impose a significant direct energy burden on the facility; however, the requirement for organic carbon to facilitate biological treatment processes to remove phosphorus (and/or nitrogen) can create a tension between these processes and any energy recovery measures that might also want to make use of this carbon.

Disinfection using chemicals does not represent a significant direct energy use, unless the energy accounting method includes "embedded energy." Ultraviolet disinfection can be a more significant energy user, as can ozone.

Tertiary and other advanced treatment, for reuse or other stringent effluent requirements, can increase the energy requirements significantly, particularly through the use of filtration technologies that have an increasing hydraulic energy requirement as the pore size is reduced, with reverse osmosis being the most energy intensive. These processes often require intermediate pumping to overcome the increased hydraulic energy requirements. Membrane bioreactors (MBRs) are a special consideration, coupling activated sludge with ultrafiltration to produce a high-quality effluent. Membrane bioreactors generally use more energy than conventional treatment because of high internal recycle flows and the need to provide scouring air to keep the membranes clean, though in recent years the energy required for scouring has been optimized and reduced significantly.

2.0 USE OF PROCESS MODELS

The design and optimization of WRRFs often uses process models, which can be a valuable tool in assessing the best approach to optimizing energy use and energy recovery. Most modern process simulators include features to track energy use. The models can be used to investigate the interaction between nutrient removal and energy recovery, through the development of a facility-wide mass balance. The Water Research Foundation (WRF) study ENER1C12 (2015) linked facility-wide models and energy calculations to produce energy flow "Sankey" diagrams to visualize energy neutrality ideas. Figure 4.2 is an example output from that study.

The WRF study investigated 25 wastewater treatment process flow schemes (configurations) that are representative of most WRRFs in North America. In addition, the project team identified modifications to specific unit processes that could be applied to certain WRRF configurations and investigated 18 "pioneering" processes that made use of current and emerging technologies to minimize energy use and/or maximize energy recovery. Some of the findings of the research are as follows:

- The full combination of best practices resulted in approximately 40% lower energy consumption than "typical" performance.
- Improving primary treatment and solids capture in thickening and dewatering processes had the most significant total positive effect of all the best practices modeled.

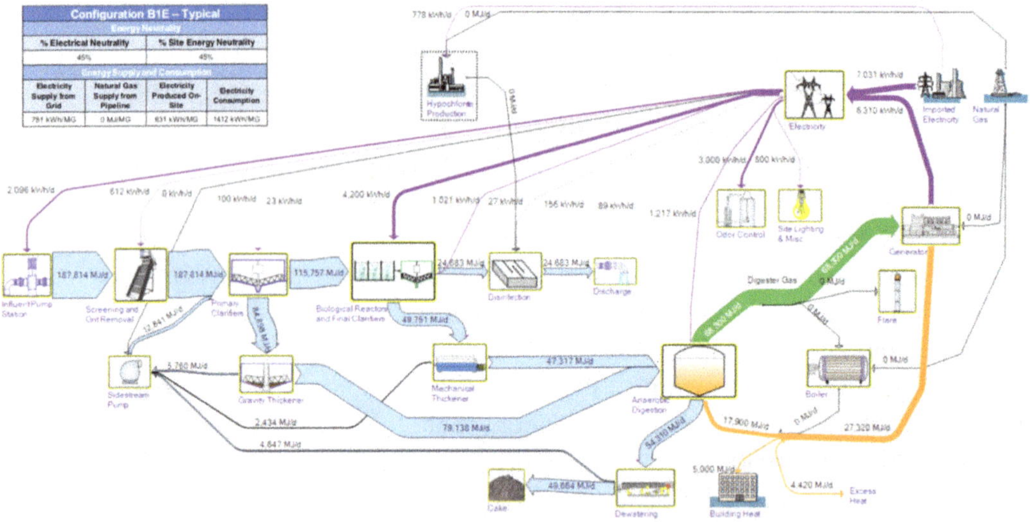

FIGURE 4.2. Example energy flow "Sankey" diagram that can be used to track different energy types facility-wide (WRF, 2015). Reprinted with permission from the Water Research Foundation.

- Significant savings in aeration blower electricity usage was achieved by reducing fouling in fine-bubble diffusers through improved operation and maintenance procedures.
- Conventional secondary treatment and nitrification facilities can be net-energy positive.
- Co-digestion of high-strength waste in anaerobic digesters was a valuable approach to achieve energy neutrality.

3.0 LIQUID PROCESSES

This section covers the liquid treatment process focusing on carbon redirection mechanisms from various physical and biological treatment technologies. In addition, this section also discusses the effect of operational parameters on carbon redirection and other process components' energy use.

3.1 Carbon Redirection

Carbon redirection technologies have been gaining attention as an important part in moving toward the net energy neutral concept in municipal WRRFs. Carbon redirection typically involves transferring of organic carbon from wastewater to sludge from which energy can then be recovered. This can be achieved through primary treatment processes and/or biological treatment processes in which biosorption plays a significant role for carbon redirection. Carbon redirection increases with decrease of solids retention time (SRT) and vice versa in activated sludge processes. After carbon redirection, the carbon-rich sludge in the liquid system requires separation from the liquid process to harvest the carbon and thus settling is critical for carbon harvesting. Many conventional primary and secondary treatment technologies have potential for capturing carbon from wastewater through innovative process implications. This section explains primary clarifier, chemically enhanced primary treatment (CEPT), primary filtration, biosorption process, and high-rate activated sludge (HRAS) technologies to understand the energy recovery mechanism.

3.1.1 Primary Clarifiers and Chemically Enhanced Primary Treatment

The primary clarifier is used in wastewater treatment because of its simple operation and low energy demand. The typical ranges of TSS, BOD, and COD removal in primary clarifiers are 50% to 70%, 25% to 40%, and 20% to 35%, respectively (Gray, 2004; Metcalf & Eddy, 2003; Water Environment Federation & Water Environment Research Foundation [WEF & WERF], 2015). The primary treatment carbon capture efficiency may be increased by adding chemicals to improve the coagulation and flocculation process between colloidal matter. The primary treatment process with chemical addition is commonly referred to as the CEPT process. Ferric chloride, aluminum compounds such as aluminum sulfate, and lime are among the popular inorganic coagulants used for CEPT application (Maktabifard et al., 2018). Besides coagulant addition, polymers (organic polyelectrolytes) as flocculent are also used to enhance the flocculation of particulate and colloidal matter. Typically, CEPT can remove 80% to 90% TSS, 50% to 80% BOD, and 45% to 80% COD (Melcer, 2003). The limitation of CEPT application is that it contains a higher percentage of additional solids, which contain high amounts of metal salts (inert compounds), which reduce the dewaterability of the primary sludge (WEF & WERF, 2015). Figure 4.3 represents typical removal efficiency of COD, TSS, and phosphorus from primary treatment processes.

3.1.2 Primary Filtration

Primary filtration is getting attention as an emerging technology for maximizing carbon redirection and harvesting and reducing aeration cost for secondary activated sludge systems. The typical TSS or volatile suspended solids, and BOD or COD removal efficiency from primary filtration are 50% to 70% and 40% to 55%, respectively (Caliskaner et al., 2017, 2018). The advantage of primary filtration technology is that it requires only 10% to 20% of the footprint of conventional primary clarifier systems, resulting

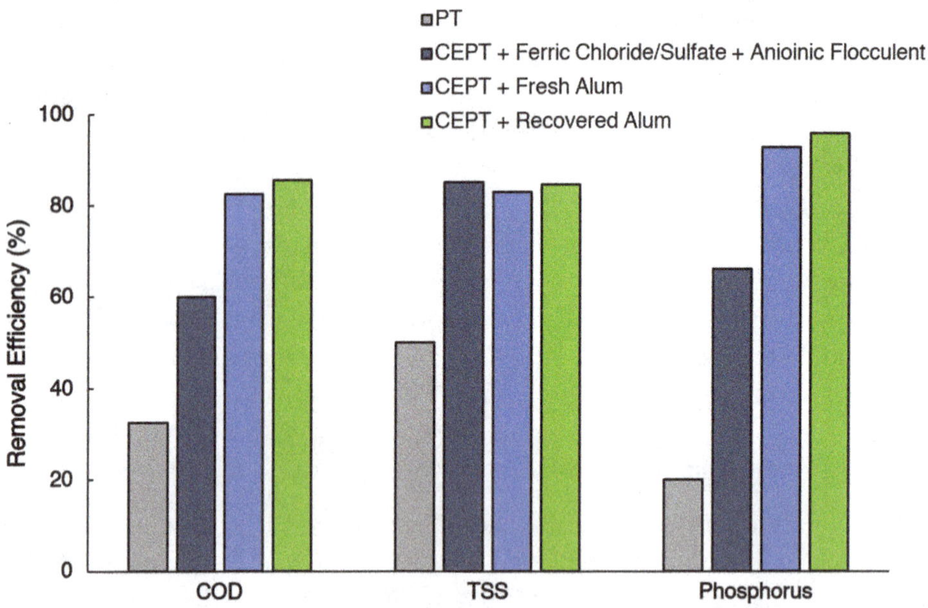

FIGURE 4.3. Typical removal efficiency of COD, TSS, and phosphorus from primary clarifier and four chemically enhanced primary treatment processes. Adapted from Maktabifard et al., 2018.

in significant capital cost savings. Because of low footprint requirement, primary filtration application can be critical technology when facility expansion is necessary by reducing the organic loading rate in secondary treatment process.

3.1.3 Biosorption Processes (High-Rate Activated Sludge)

In this section, the *biosorption* is referred as extracellular adsorption (or sorption) and intracellular storage and biosorption is one of key mechanisms for carbon redirection in biological treatment processes. Adsorption and entrapment or enmeshment are not distinct mechanisms because these depend on extracellular polymeric substances (EPS) and thus they are the same phenomena (Rahman et al., 2017). The adsorption mechanism includes "outer cell surface adsorption of particulate and colloidal organic matters," which is a physical-chemical based phenomenon between activated sludge flocs and organic matter in wastewater (Guellil et al., 2001; Majone et al., 1999; Modin et al., 2016). The adsorption mechanism is assumed to entrap both biodegradable and unbiodegradable (inert) organic compounds from wastewater using EPS excretion at outside of cell surface. The adsorption is thus considered as the dominant mechanisms to remove COD in HRAS processes. This is because the mechanism removes organic matter by a rapid response and can reach the equilibrium within a minute (Guellil et al., 2001).

High-rate activated sludge systems can maximize carbon redirection and harvesting from wastewater through biosorption. The operational parameters for HRAS systems are reported over a wide range of values in literature, which makes the definition of HRAS systems confusing. Metcalf and Eddy (2003) reported the HRAS systems operational SRT and hydraulic residence time (HRT) are 0.5 to 2 days and 1.5 to 3 hours, respectively. Orhon (2015) explained the HRAS system's SRT range are to be 2 to 4 days and HRT of 1 to 3 hours. Gray (2004) proposed the range of SRT for HRAS system should be 0.2 to 0.5 days, which is based on A-stage systems and HRT of 1 to 2 hours. The volumetric BOD loading rate is also reported in wide spectrum of 1.2 to 6 kgBOD/m^3·d. The energy savings by minimizing carbon oxidation and carbon recovery from wastewater is the main benefit of HRAS systems in modern activated sludge design.

The A-stage of an adsorption/bio-oxidation process is a well-known HRAS system used in Europe that intends to minimize carbon oxidation and maximize carbon harvesting through biosorption at a low SRT (0.1 to 0.6 days) conditions (Boehnke et al., 1998; de Graaff et al., 2016; Haider et al., 2003; Kinyua et al., 2017; Miller et al., 2017; Rahman et al., 2019; Smitshuijzen et al., 2016).

Within the wide spectrum of activated sludge processes, high-rate contact stabilization is a variation of the conventional activated sludge process that aims to select for a strong biosorption response of the sludge and, thus, further improve the overall carbon harvesting (capture) from wastewater (Rahman et al., 2016). However, the application of contact stabilization process for wastewater treatment has gradually decreased in the United States. Few contact stabilization systems have been implemented as HRAS systems to take advantage of the fundamental mechanism (return activated sludge [RAS] aeration) to maximize the carbon redirection and harvesting from wastewater with balanced carbon management and inventory level (Dolejs et al., 2016; Khararjian & Sherrard, 1978; Meerburg et al., 2016).

3.1.4 Combining Carbon Diversion Technologies

Carbon diversion and harvesting technology has been investigated through physical treatment, biological treatment, and combination of physical and biological treatment processes. The combination of primary treatment or CEPT and high-rate contact stabilization has been studied successfully to recover carbon from wastewater (Rahman et al., 2016, 2017). A recent technology innovation comprised of low dissolved oxygen contact tanks integrated with dissolved air flotation systems has been applied before secondary activated sludge systems (Waul et al, 2016). Other carbon harvesting technology such as high-rate contact stabilization integrated with MBR or

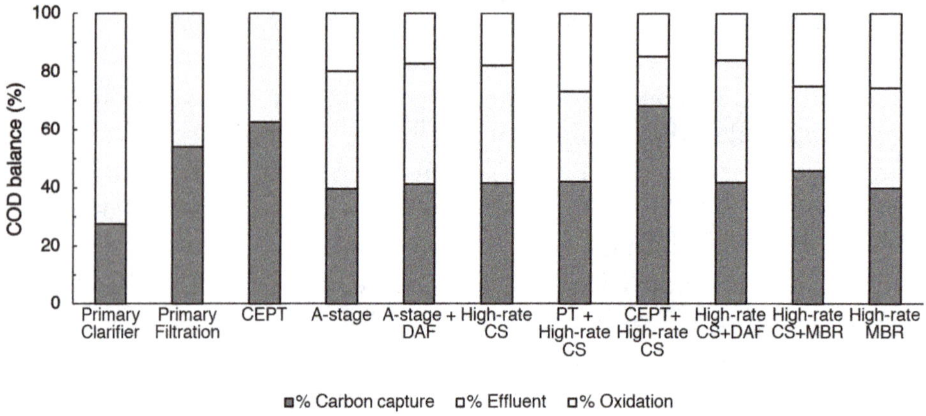

FIGURE 4.4. Typical carbon balance of different carbon harvesting technologies treating wastewater

high-rate MBR systems are getting attention as alternative carbon redirection processes (Dai et al., 2018). Figure 4.4 represents the summary of typical carbon balance (as % carbon harvesting, % carbon oxidation, and % effluent COD) for various carbon capture technologies.

The carbon balance indicates that oxidation generally accounts for 16% to 27% of influent COD in any high-rate biological treatment processes and this is critical for biomass growth resulted in flocculation ability. The ability of carbon capture in high-rate processes ranges between 40% and 60% except when CEPT integrates with high-rate contact stabilization technology (68%). Using physical treatment process, CEPT and primary filtration are competitive compared to biological treatment process because these systems can harvest carbon between 54% and 63%. Thus, the selection of technology is dependent on existing infrastructure of WRRF and their treatment goals.

3.1.5 Anaerobic Treatment and Dissolved Methane Challenges

The implication of anaerobic treatment technologies such as anaerobic membrane bioreactor (AnMBR) and upflow anaerobic sludge blanket digestion in domestic wastewater treatment are gaining attention because of their low energy consumption. Among several anaerobic treatment processes, AnMBR technology is an emerging technology that has been applied successfully for industrial wastewater treatment that has a high concentration of soluble COD and produces more methane. The economic efficiency of AnMBR for domestic wastewater treatment is also promising. The principal driver of implementing full-scale AnMBR is to achieve energy neutral wastewater treatment through a reduction in net energy demand by increasing the total biogas productions. The AnMBR can either be used for primary effluent or for raw wastewater.

Literature suggested that the typical range of headspace methane production in AnMBR is 0.02 to 0.27 L CH_4/g COD treating primary effluent or raw wastewater (Guo et al., 2016; Wang et al., 2018). However, the stoichiometric value for methane production is 0.35 L CH_4/g COD; this is because the primary effluent contains 50% of the dissolved methane (Velasco et al., 2018). An AnMBR is favorable under mesophilic and thermophilic conditions for improved economic efficiency but recovery of methane dissolved in the primary effluent is a significant hurdle in its widescale application.

3.2 Summary and Overview of Liquid Process Energy Use

The liquid treatment processes of a WRRF typically consume more than 70% of total electricity in which aeration is one of the major energy consumers. A typical electricity distribution in different liquid and solids treatment process area are shown in Figure 4.5.

This section will discuss the effect of operational parameters (especially on HRAS systems), and energy saving mechanism for pumping, aeration, passive aeration, filtration, and disinfection processes applicable to liquid treatment.

3.2.1 Operational Parameters

The SRT, HRT, dissolved oxygen, recycle ratio, organic loading rate, and temperature have significant influence on the biochemical transformation processes, which relate to the microorganism's growth rate especially in HRAS systems that maximizes carbon redirection and harvesting.

By changing the SRT, one can have a significant influence on bioflocculation capacity, sludge physiochemical properties (hydrophobicity and surface charge) that controls the flocculation capacity, EPS production, and even for removing micropollutants (Clara et al., 2005; Liao et al., 2001; Sesay et al., 2006).

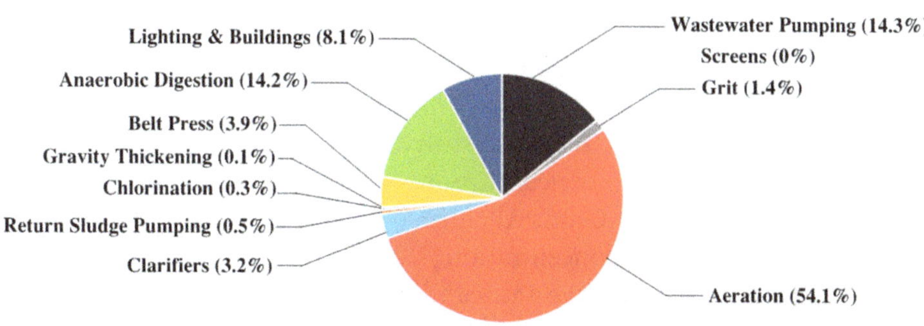

FIGURE 4.5. Typical electricity requirements for activated sludge WRRF. Adapted from WEF, 2009.

Several studies also confirmed that bioflocculation and sedimentation capability and EPS production have a strong correlation with SRT (Faust et al., 2014a and b; Jimenez et al., 2015; Li & Yang, 2007). Saunders and Dick (1981) reported that SRT has a negative correlation with effluent soluble COD concentration and indicated that at low SRT conditions (< 0.5 day), the A-stage discharges significant fraction of biodegradable COD through effluent (Haider et al., 2003). The SRT also has a significant effect on carbon oxidation. The higher the SRT in a system, the higher the carbon oxidation, and vice versa for short SRT systems (Ge et al., 2013; Jimenez et al., 2015). In terms of energy perspective, if SRT is low compared to conventional systems, the aeration cost can be significantly reduced using HRAS systems.

Dissolved oxygen concentration is one of the important parameters for organic removal, bacterial metabolism, and settleability. At low dissolved oxygen concentrations, it may have a negative effect on the sludge settleability and the metabolic activity of microorganisms, which may eventually generate an incomplete COD removal in the systems (Faust et al., 2014a and b; Jimenez et al., 2015; Kinyua et al., 2017; Zhang & Grant Allen, 2008). For energy savings perspective, the goal is to find the optimum dissolved oxygen level for the systems that can be implemented by balancing metabolic activity, carbon removals, and settleability. Jimenez et al. (2015) observed that dissolved oxygen has a significant role on particulate, colloidal, and soluble COD removal efficiency. In their study, at dissolved oxygen < 0.5 mg O_2/L and HRAS system operated at 1-day SRT and 30 minutes of HRT, the removal efficiency dramatically dropped and thus effluent solids concentration deteriorated (Zhang & Grant Allen, 2008). The influence of dissolved oxygen on EPS production was also reported in that study and found that EPS decreased with decreasing dissolved oxygen concentration. The critical dissolved oxygen was 0.5 mg O_2/L, below which EPS production dropped significantly. Faust et al. (2014a and b) also observed the effect of dissolved oxygen on EPS production in high-rate MBR systems operated at 0.5-day SRT in which the bound EPS was higher at dissolved oxygen of 4 mg O_2/L than dissolved oxygen of 1 mg O_2/L condition. In addition, the flocculation of colloidal COD and settleability was better at higher dissolved oxygen conditions compared to low dissolved oxygen concentration. Similar observations were also examined by Zhang and Grant Allen (2008) on activated sludge systems. At low dissolved oxygen concentration (<1 mg O_2/L), the dissolved oxygen had negative effect on sludge settleability concentrations and promotes filamentous growth (Zhang & Grant Allen, 2008).

3.2.2 Pumping

Wastewater pumping is one of the major components that accounts 14% of the total energy used to operate the facility. The most important factor in providing efficient pumping is to match the pump size to the required flow and pressure; that is, sizing the pump correctly for the most common duty. Where a pump is required to deliver a range of flows, one of the major steps to save energy for pumping system is to use variable frequency drives (VFDs) on pumps. Adjusting the pump speed with varying flowrate can reduce energy consumption significantly especially when pump speed can be decreased instead of using constant speed (Idaho Power, 2018). Besides pump speed strategy, it is always better to replace standard efficiency motors with premium efficiency motors that can minimize the energy consumption (Idaho Power, 2018). In addition, it is always better to replace worn or inefficient pumps with high-efficiency pumps. Determining the pump system efficiency over a range of pumping flow requirements always optimizes the energy usage. More details on pumping are covered in Chapter 5.

3.2.3 Aeration

Aeration is one of the highest energy consumers in a WRRF and typically uses 40% to 60% of total energy consumed in a WRRF. There are several ways to save energy from aeration if proper steps can be taken. Replacing coarse-bubble diffusers with fine-bubble diffusers is one of the critical pathways to improve oxygen transfer efficiency. The aeration blower load can be reduced by automatic control of dissolved oxygen through installation of dissolved oxygen sensors in tanks integrated with a control system. In aeration zones where high dissolved oxygen is required, implementation of most open valve strategies will reduce the pressure of blowers resulting in reduced energy consumption. The dissolved oxygen can be controlled in remaining aeration zones, which can be controlled by valves that can maintain the dissolved oxygen setpoint for each zone. Other sensor-based advanced aeration control systems such as ammonia-based aeration control and ammonia versus nitrate control strategies have recently gained interest for reducing aeration energy cost (Maktabifard et al., 2018). In terms of passive aeration, trickling filter, rotating biological contactor, and cascade aeration are among the major process components in which aeration control can be implemented. Chapter 7 has more details on aeration energy.

Table 4.1 describes the summary of different liquid treatment process components (pumping, aeration, filtration, and disinfection) in which energy consumption can be optimized through different operational and maintenance steps.

TABLE 4.1. Summary of different process optimization techniques to minimize energy consumption. Adapted from Idaho Power, 2018; Energy Trust Oregon, 2017; New York State Energy Research & Development Authority, 2010; Cortez et al., 2008.

Process Components	Process Area	Process Technology	Best Practice	Potential Energy Savings (%)
Pumping (Influent/ RAS/WAS/Aeration Basin Recirculation)	All pumping applications	Optimize pump system efficiency	Run the pump at optimum operational conditions (startup flow and design flow capacity)	15–30
	Booster pump discharges, tank filling	Avoid pump discharge throttling	Avoid using throttling of valve to control flow, use VFD	>15–50
	Aeration blower, disinfection, clarifiers, pumps	High-efficiency motors	Replace low-efficiency motor with high-efficiency motor	5–10
Optimize aeration systems	Aeration tanks, aerobic digester	Fine-bubble aeration	Replace coarse-bubble with fine-bubble diffuser, clean the systems periodically	20–75
	Aeration tanks, aerobic digester	Air blower on VFD	Variable air supply capability, avoid air-flow discharge throttling	15–50
	Aeration tanks, aerobic digester, post-aeration	Dissolved oxygen control	Monitor and control dissolved oxygen by varying air-flow rate	20–50
	Post-aeration of facility effluents	Cascade aeration*	Replace submerged diffuser system with cascade aeration	100
	Trickling filter	Secondary treatment	Optimum plastic media (cross-flow) provides better oxygen transfer	25
	Rotating biological contactor	Dissolved oxygen control	Monitor and control dissolved oxygen by varying airflow rate	75
Filtration system	Tertiary treatment	Cloth media filter	Sequencing backwash cycles, off-peak backwash times	Minor
Disinfection	Tertiary treatment	Ultraviolet system	Replace medium pressure lamp with low-pressure high output lamp	50
			Regular cleaning of lamp will prevent lamp sleeve fouling	10

*Cascade aeration only works if the WRRF has available head in their hydraulics.

4.0 SOLIDS PROCESSING

4.1 Biogas Production

Biogas, or digester gas, is produced in anaerobic digesters. The gas is approximately 60% methane, the same fuel as in natural gas. Facilities with anaerobic digesters typically use a flare to burn excess methane gas. Small facilities can use the gas to fuel the boiler(s) used to heat the incoming sludge, although a special burner is needed to do so. Larger facilities use the gas to fuel engines that power pumps, aeration blowers, or engine generators. The fuel has also been cleaned and discharged to natural gas pipelines. For further information refer to Chapter 10.

4.1.1 Co-Digestion to Boost Production

Co-digestion is the practice of adding compatible high strength wastes to an anaerobic digester that is also digesting wastewater solids. Co-digestion provides a number of benefits. It increases the production of digester gas, which can be used as a fuel. Co-digestion also keeps the high-strength waste out of the liquid stream, reducing the demand for power to provide aeration in biological treatment. One negative effect of co-digestion—depending on the feed material—is that it can introduce nitrogen and/or phosphorus to the system, some of which will be returned to the liquid treatment in sidestreams. For further information refer to Chapter 10.

4.2 Solids Processing Energy Use

4.2.1 Thickening

Thickening is used to concentrate solids for further processing. Thickening the solids reduces the volume of sludge to be handled, and thereby the size required for downstream processes. Thickening may also reduce chemicals required for dewatering. Digester performance in both aerobic and anaerobic systems is dependent on SRT, with longer SRTs providing enhanced stabilization and pathogen reduction. Thickening of solids is important for digestion because the process

- decreases the mass of liquid to heat in anaerobic digestion,
- may allow auto-heating of the digester contents in autothermal thermophilic aerobic digestion systems,
- affects mixing processes and oxygen transfer processes, and
- increases detention time of digesters.

Gravity thickeners are the lowest energy users for thickening but should be operated to provide optimal concentration of solids without allowing needless aging of the solids or excessive loss or washout of solids in the recycle stream. Belt and drum thickeners have a higher energy use and should be operated to provide a high solids capture rate while achieving desired solids concentration. Capture rates of 90% or more are achievable and should be targeted even at the expense of slightly less-thickened solids.

Thickening centrifuges are also available. They have a higher energy demand but may reduce chemical conditioning costs and should be considered in life-cycle analysis to assess which method of thickening provides the lowest cost to own and operate.

4.2.2 Dewatering

Dewatering is used to concentrate wastewater solids to 16% to 25% solids, depending on the type of sludge or biosolids. Dewatering is used to remove water and reduce the volume of material for handling, further processing, transport, and storage.

For example, in facilities that use thermal processes such as driers or incineration, good dewatering is critical in reducing the water content of the sludge. Modern fluidized-bed incinerators can be autogenous (i.e., requiring no external fuel) if the feed cake is sufficiently dry.

Dewatering commonly involves the addition of a polymer to agglomerate solids into larger particles followed by dewatering in a belt filter or screw press. Another option is the use of a dewatering centrifuge.

The belt filter press and screw press dewatering options described above are relatively low energy intensity. Centrifuges require additional horsepower, but typically produce a drier cake. If the design criteria require this drier cake, then centrifuges may be the only choice. Otherwise, a life-cycle cost analysis is required to determine if haul costs offset the increased power costs. Another option for dewatering is a plate and frame press, though this older technology is less common.

An often overlooked but important aspect of dewatering is solids recovery. If solids recovery is poor, the solids are recycled to the treatment process. With continued poor recovery, the solids in the recycle stream will continue to grow. This requires additional dewatering time, additional pumping, and may have a negative effect on other parts of the treatment process.

4.2.3 Sludge Pumping

Sludge pumping covers a number of different applications. Pumping unthickened waste activated sludge (WAS) and thin primary sludge is similar to pumping water. However, thickened primary sludge and thickened WAS becomes increasingly viscous as the percentage solids increases and behaves differently than water. There are some applications of pumping dewatered sludge cake that has different pumping characteristics. It is important to work with vendors to select the correct pump for the application.

There are innovative impeller designs such as the screw centrifugal pump that have demonstrated ability to pump at lower power requirements that can be explored as an energy conservation method.

Having adequate power capacity (large enough motor) and pressure rating are two important parameters to ensure satisfactory performance.

There are applications in which installation of a grinder ahead of the pump may reduce pumping power.

Refer to Chapter 5 for more details on pumping in general.

5.0 OTHER ENERGY DEMANDS

5.1 Process Heating—Digesters, Including Thermal Hydrolysis Process

The principal process heating requirement is for anaerobic digestion of the wastewater solids. Digesting solids must be maintained at the correct temperature (e.g., 35.0 to 36.7 °C [95 to 98 °F] for mesophilic anaerobic digestion), and raw sludge must be heated to the digester temperature.

Increasing interest in thermal hydrolysis processes provide another potential point of demand for process heating.

Another process heating requirement is preheating of feed air to an incinerator. Process water may also be used to cool/quench the ash from an incinerator.

Facilities that have anaerobic digesters can provide the heating requirements with digester gas produced during the digestion process. This gas can also be used in cogeneration for other energy uses such as heating, ventilation, and air conditioning. Another potential source is solar thermal energy. More details on heat and energy around digestion is given in Chapter 10. Details on renewable energy is given in Chapter 11.

5.2 Ventilation and Odor Control

Energy is also used for air-handling units for ventilation, and odor control units. There are odor control units with 25-, 50- and 75-HP motors that operate 24 hours per day, 7 days per week, 52 weeks per year. Making sure that the motors are maintained, use of premium efficiency motors, and use of VFDs can all contribute to optimizing energy use for these units.

In facilities in which many processes are housed within buildings, ventilation flowrates can be significantly higher and cause the energy use for this to be a significant contribution to the overall facility energy. For covered basins, limiting the volume of headspace is a good way to reduce the amount of air that has to be moved and treated.

5.3 Building Heating and Cooling

Wastewater treatment facilities have buildings for housing equipment and personnel for facility operation, maintenance, laboratory, and administration. All of these buildings use energy for space heating during colder weather in cold climates and air-conditioning during warmer weather. There are a surprising number of makeup air units and exhaust fans powered with motors ranging in size from 25 to 75 HP. These units are often overlooked, and it is important that they receive periodic maintenance to make sure that belt drives are in good operating condition.

5.4 Nonpotable Water System

Examples are seal water, service water, and, sometimes, potable water. Each of these systems may operate a pump all the time to maintain system pressure. These systems may include a blow-off valve or recirculation line for times when flow is low. A better option is to install a VFD and to vary the speed of the pump based on system pressure. Another option is installation of a pressure tank to maintain system pressure and potentially refilling during off-peak times.

5.5 Lighting

Energy is required for the lighting of work areas, outside areas, and interconnecting gallery areas. Lighting technology continues to improve providing better illumination with less power and the use of photocells to turn lights on and off as needed. Any facility could benefit from a lighting survey by a lighting contractor to learn about opportunities for improvement. The benefit to hiring a contractor is that all of the units are updated at the same time, maximizing savings.

5.6 Other Uses

Other energy uses that can be considered are such things as auxiliary service compressed air, instrument air, hoisting cranes, and sump pumps.

Some minor energy uses include instrumentation, electric-operated valves, gas pilots for boilers, digester gas flares, unit heaters, and portable steam units.

6.0 REFERENCES

Boehnke, B., Schulze-Rettmer, R., & Zuckut, S. (1998). Cost-effective reduction of high-strength wastewater by adsorption-based activated sludge technology. *Water Engineering and Management*, *145*, 31–34.

Caliskaner, O., Tchobanoglous, G., Reid, T., Davis, B., Young, R., & Downey, M. (2017). First full-scale installation of primary filtration for advanced primary treatment to save energy and increase capacity. *Proceedings of the Water Environment Federation*, *2017*(7), 4241–4254.

Caliskaner, O., Tchobanoglous, G., Wu, Z., Young, R., Paez, C., Davis, B., Mansell, B., Ackman, P., Reid, T., & Dyson, J. (2018). Performance of full scale and demonstration-scale primary filtration projects. *Proceedings of the Water Environment Federation*, *2018*(18), 70–84.

Clara, M., Kreuzinger, N., Strenn, B., Gans, O., & Kroiss, H. (2005). The solids retention time—A suitable design parameter to evaluate the capacity of wastewater treatment plants to remove micropollutants. *Water Research*, *39*(1), 97–106.

Cortez, S., Teixeira, P., Oliveira, R., & Mota, M. (2008). Rotating biological contactors: A review on main factors affecting performance. *Reviews in Environmental Science and Biotechnology*, *7*(2), 155–172.

Dai, W., Xu, X., & Yang, F. (2018). High-rate contact stabilization process-coupled membrane bioreactor for maximal recovery of organics from municipal wastewater. *Water*, *10*(7), 878.

de Graaff, M. S., van den Brand, T. P., Roest, K., Zandvoort, M. H., Duin, O., & van Loosdrecht, M. C. (2016). Full-scale highly-loaded wastewater treatment processes (A-Stage) to increase energy production from wastewater: performance and design guidelines. *Environmental Engineering Science*, *33*(8), 571–577.

Dolejs, P., Gotvald, R., Velazquez, A. M., Hejnic, J., Jenicek, P., & Bartacek, J. (2016). Contact stabilization with enhanced accumulation process for energy recovery from sewage. *Environmental Engineering Science*, *33*(11), 873–881.

Energy Trust of Oregon. (2017). *Water and wastewater treatment—Energy Savings Guide.*

Faust, L., Temmink, H., Zwijnenburg, A., Kemperman, A., & Rijnaarts, H. (2014a). Effect of dissolved oxygen concentration on the bioflocculation process in high loaded MBRs. *Water Research*, *66*, 199–207.

Faust, L., Temmink, H., Zwijnenburg, A., Kemperman, A., & Rijnaarts, H. (2014b). High loaded MBRs for organic matter recovery from sewage:

Effect of solids retention time on bioflocculation and on the role of extracellular polymers. *Water Research*, *56*, 258–266.

Ge, H., Batstone, D. J., & Keller, J. (2013). Operating aerobic wastewater treatment at very short sludge ages enables treatment and energy recovery through anaerobic sludge digestion. *Water Research*, *47*(17), 6546–6557.

Gray, N. F. (2004). *Biology of wastewater treatment*. World Scientific.

Guellil, A., Thomas, F., Block, J.-C., Bersillon, J.-L., & Ginestet, P. (2001). Transfer of organic matter between wastewater and activated sludge flocs. *Water Research*, *35*(1), 143–150.

Guo, W., Ngo, H. H., Chen, C., Pandey, A., Tung, K. L., & Lee, D. J. (2016). Anaerobic membrane bioreactors for future green bioprocesses. *Green Technologies for Sustainable Water Management, ASCE*. https://doi.org/10.1061/9780784414422.ch25

Haider, S., Svardal, K., Vanrolleghem, P., & Kroiss, H. (2003). The effect of low sludge age on wastewater fractionation. *Water Science & Technology*, *47*(11), 203–209.

Idaho Power. (2018). *Eliminating energy waste from wastewater: Wastewater energy efficiency cohort success story and energy-saving tips*.

Jimenez, J., Miller, M., Bott, C., Murthy, S., De Clippeleir, H., & Wett, B. (2015). High-rate activated sludge system for carbon management—Evaluation of crucial process mechanisms and design parameters. *Water Research*, *87*, 476–482.

Khararjian, H. A., & Sherrard, J. H. (1978). Contact stabilization treatment of a colloidal organic wastewater. *Journal (Water Pollution Control Federation)*, 645–652.

Kinyua, M. N., Elliott, M., Wett, B., Murthy, S., Chandran, K., & Bott, C. B. (2017). The role of extracellular polymeric substances on carbon capture in a high rate activated sludge A-stage system. *Chemical Engineering Journal*, *322*, 428–434.

Li, X., & Yang, S. (2007). Influence of loosely bound extracellular polymeric substances (EPS) on the flocculation, sedimentation and dewaterability of activated sludge. *Water Research*, *41*(5), 1022–1030.

Liao, B., Allen, D., Droppo, I., Leppard, G., & Liss, S. (2001). Surface properties of sludge and their role in bioflocculation and settleability. *Water Research*, *35*(2), 339–350.

Majone, M., Dircks, K., & Beun, J. J. (1999). Aerobic storage under dynamic conditions in activated sludge processes. The state of the art. *Water Science and Technology*, *39*(1), 61–73.

Maktabifard, M., Zaborowska, E., & Makinia, J. (2018). Achieving energy neutrality in wastewater treatment plants through energy savings and enhancing renewable energy production. *Reviews in Environmental Science and Bio/Technology*, *17*(4), 655–689.

Meerburg, F. A., Boon, N., Van Winckel, T., Pauwels, K. T. G., & Vlaeminck, S. E. (2016). Live fast, die young: Optimizing retention times in high-rate contact stabilization for maximal recovery of organics from wastewater. *Environmental Science & Technology*, *50*, 9781–9790.

Melcer, H. (2003). *Methods for wastewater characterization in activated sludge modeling*. IWA Publishing.

Metcalf & Eddy. (2003). *Wastewater engineering: Treatment and reuse*. McGraw Hill.

Miller, M. W., Elliott, M., DeArmond, J., Kinyua, M., Wett, B., Murthy, S., & Bott, C. B. (2017). Controlling the COD removal of an A-stage pilot study with instrumentation and automatic process control. *Water Science and Technology*, *75*(11-12), 2669–2679.

Modin, O., Persson, F., Wilén, B. M., & Hermansson, M. (2016). Nonoxidative removal of organics in the activated sludge process. *Critical Reviews in Environmental Science and Technology*, *46*(7), 635–672.

New York State Energy Research & Development Authority. (2010). *Water and wastewater energy management. Best practices handbook.*

Orhon, D. (2015). Evolution of the activated sludge process: The first 50 years. *Journal of Chemical Technology and Biotechnology*, *90*(4), 608–640.

Rahman, A., Meerburg, F. A., Ravadagundhi, S., Wett, B., Jimenez, J. A., Bott, C., Al- Omari, A., Riffat, R., Murthy, S., & De Clippeleir, H. (2016). Bioflocculation management through high-rate contact-stabilization: A promising technology to recover organic carbon from low-strength wastewater. *Water Research*, *104*, 485–496.

Rahman, A., Mosquera, M., Thomas, W., Jimenez, J. A., Bott, C., Wett, B., Al-Omari, A., Murthy, S., Riffat, R., & De Clippeleir, H. (2017). Impact of aerobic famine and feast condition on extracellular polymeric substance production in high-rate contact stabilization systems. *Chemical Engineering Journal*, *328*, 74–86.

Rahman, A., De Clippeleir, H., Thomas, W., Jimenez, J. A., Wett, B., Al-Omari, A., Murthy, S., Riffat, R., & Bott, C. (2019). A-stage and high-rate contact-stabilization performance comparison for carbon and nutrient redirection from high-strength municipal wastewater. *Chemical Engineering Journal*, *357*, 737–749.

Saunders, F. M., & Dick, R. I. (1981). Effect of mean-cell residence time on organic composition of activated sludge effluents. *Journal (Water Pollution Control Federation)*, *53*(2), 201–215.

Sesay, M. L., Özcengiz, G., & Sanin, F. D. (2006). Enzymatic extraction of activated sludge extracellular polymers and implications on bioflocculation. *Water Research*, *40*(7), 1359–1366.

Smitshuijzen, J., Pérez, J., Duin, O., & van Loosdrecht, M. C. (2016). A simple model to describe the performance of highly-loaded aerobic COD removal reactors. *Biochemical Engineering Journal*, *112*, 94–102.

Velasco, P., Jegatheesan, V., & Othman, M. (2018). Recovery of dissolved methane from anaerobic membrane bioreactor using degassing membrane contactors. *Frontiers in Environmental Science*, *6*, 151.

Wang, K. M., Martin Garcia, N., Soares, A., Jefferson, B., & McAdam, E. J. (2018). Comparison of fouling between aerobic and anaerobic MBR treating municipal wastewater. *H_2O Open Journal*, *1*(2), 131–159.

Waul, C., Doyle, M., Smith, G., Pino-Jelcic, S., & Erdogan, A. (2016). Shifting the energy balance with biologically enhanced primary treatment—How carbon diversion makes sense. *Proceedings of the Water Environment Federation*, *2016*(3), 945–960.

Water Environment Federation. (2009). *Energy conservation in water and wastewater treatment facilities* (WEF Manual of Practice No. 32). Water Environment Federation.

Water Environment Federation & Water Environment Research Foundation. (2015). *Shortcut nitrogen removal—Nitrite shunt and deammonification.* Water Environment Federation.

Water Research Foundation. (2015). *A guide to net-zero energy solutions for water resource recovery facilities* (Final report from Project ENER1C12).

Zhang, Y., & Grant Allen, D. (2008). The effect of short-term dissolved oxygen transients on activated sludge. *Water Quality Research Journal Canada*, *43*(2-3), 201–210.

5

Pumping Systems

Ned Paschke, PE; Rachel Feil; Matt Seib, PhD, PE; & Eric Hjellen

1.0 INTRODUCTION

Pumping is an essential process for almost all water resource recovery facilities (WRRFs) and wastewater collection systems. Pumps of many different types and sizes are required for wastewater pumping stations, influent lift stations, and treatment facility processes.

The energy used and power required for pumping can be quite substantial. In many WRRF systems, pumping represents the second largest user of energy, after aeration. In systems with high pumping needs, pumping can be the largest energy user.

This chapter examines the key parameters that influence input power and energy use in pumping processes for WRRFs and collection systems and will discuss opportunities to reduce pumping power and energy use.

The water resource recovery industry requires a special emphasis on robustness and reliability, and energy is only one of many important considerations. A particular energy-saving practice would not be wise if it saves energy at the expense of reliability, or if it is unnecessarily complex or costly. Fortunately, there are multiple opportunities to save energy and power without adversely affecting the facility or process.

2.0 INPUT POWER AND ENERGY USED FOR PUMPING—GENERAL CONCEPTS

The power and energy required for pumping depend on four important parameters: flowrate, pumping head, equipment efficiencies, and elapsed time. The combination of these four parameters will determine the energy used by a particular pumping system over a time period, as well as the power drawn by that system at any particular point in time.

The examples and discussions in this chapter consider that a pumping unit consists of a pump, an electric motor, and (in some cases) a variable frequency drive (VFD). Also, it is assumed that the wastewater being pumped is an essentially incompressible fluid with a density close to 1.0 kg/L. With these conditions, the power and the energy required for the steady-state pumping of water or wastewater can be expressed as follows:

$$\text{Water power}_{\text{kW}} = \left[\frac{1}{1000}\right] Q\rho g \Delta H_p = 0.00981 Q\rho g \Delta H_p \tag{5.1}$$

$$\text{Input power to pump}_{\text{kW}} = 0.00981 \frac{Q\Delta H_p}{e_p} \tag{5.2}$$

$$\text{Input power to motor}_{\text{kW}} = 0.00981\frac{Q\Delta H_p}{e_p e_m} \tag{5.3}$$

$$\text{Input power to system}_{\text{kW}} = 0.00981\frac{Q\Delta H_p}{e_p e_m e_{vfd}} \tag{5.4}$$

$$\text{Energy used by system kWhr} = 0.00981 Q\Delta H_p e_p e_m e_{vfd\Delta t} \tag{5.5}$$

Where

Water power $_{\text{kW}}$ = instantaneous power, in kilowatts, being produced by the pump and added to the flowing water;

Input power to pump $_{\text{kW}}$ = instantaneous power, in kilowatts, being provided by the motor, and flowing into the pump;

Input power to motor $_{\text{kW}}$ = instantaneous power, in kilowatts, flowing into the motor;

Input power to system $_{\text{kW}}$ = instantaneous power, in kilowatts, flowing into the overall pumping system (pump + motor + VFD). Where no VFD is used, this is the same as input power to the motor;

Energy used by system $_{\text{kWhr}}$ = energy used by the pumping system, in kilowatt hours, over a time period Δt in hours;

Q = the flowrate, in liters per second, of water or wastewater being pumped at a particular point in time;

ΔH_p = the pumping head being produced by the pump, in meters, at a particular point in time. The pumping head (also called the TDH, or total differential head), is the difference in head across the pump, from immediate suction side to discharge side;

ρ = density of water or wastewater = 1 kg/L;

g = acceleration as a result of gravity = 9.81 m/s^2;

Δt = elapsed run time of the pump (in hours);

e_p = the efficiency of the pump alone (not including the motor);

e_m = the efficiency of the electric motor driving the pump; and

e_{vfd} = the efficiency of the VFD supplying power to the electric motor. If no VFD is used, the effective e_{vfd} = 1.00.

Note: all efficiencies (e_p, e_m, e_{vfd}) are dimensionless, and expressed in decimal form.

The coefficient 0.00981 above is for use of the International Standard units shown. It can be revised as needed for other units of flow, head, and time.

For example, if flow is expressed in gallons per minute (gpm) and if pumping head is expressed in feet, the equations would be revised as follows:

$$\text{Input power to system}_{\text{kW}} = 0.00189 \frac{Q\Delta H_p}{e_p e_m e_{vfd}} \tag{5.6}$$

$$\text{Energy used by system kWhr} = 0.00189 Q\Delta H_p e_p e_m e_{vfd\Delta t} \tag{5.7}$$

Where

Q = flowrate, in gallons per minute;
ΔH_p = pumping head, in feet; and
Δt = elapsed run time of the pump, in hours.

All other terms are the same as defined previously

In a pumping operation, flow, pumping head, elapsed time, and equipment efficiencies are not fully independent of each other. When one parameter is changed, it often creates a change in the others. Therefore, when considering how the energy use or the input power will be affected by a proposed change (for example, a change in pump size, pump speed, pipe size, control elevations, etc.), the combined changes in all four parameters must be considered. The example calculations presented later in this chapter will help to illustrate this.

Note that a motor produces only as much power as *required* by the hydraulic operating parameters (flow, pumping head, and pump efficiency), as shown in the above equations. Although a motor will have a nominal nameplate rating (for example, 74.6 kW or 100 HP), this does not mean that the motor produces the full 74.6 kW or 100 HP when it is operating. It only means that it is *capable* of producing that much power if required. In effect, the pumping process "pulls" from the motor only as much power as needed to drive that particular pump impeller to deliver that flowrate in that particular piping system.

3.0 VARIABILITY OF FLOWRATES

Most pumping facilities in WRRF systems must be prepared to handle a wide variety of incoming flowrates. This is particularly true in collection systems, but also within treatment facilities. This wide range of hydraulic loading has a large effect on the facility's physical dimensions, costs, pump selections, operating strategies, and power and energy use.

A facility typically experiences a diurnal flow pattern (such as in Figure 5.1), with lowest flows occurring during late night/early morning, and

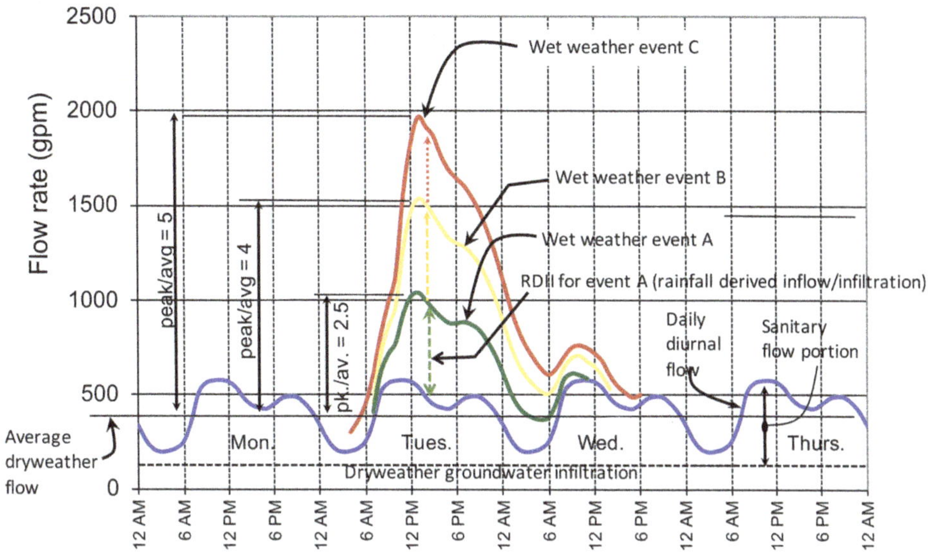

FIGURE 5.1. Example of wastewater flow variations

higher flows occurring in mid-day/early evening. While these daily ranges can be significant, they are often far exceeded during wet weather events (Figure 5.1). Large increases in flow can occur in separate sanitary sewer systems because of leakage of stormwater and groundwater (inflow and infiltration), and even larger flow increases can occur in combined sewer systems (which directly receive the inflow of stormwater). The incoming peak flow at a WRRF depends on the size and scope of the wet weather event, the type and condition of the collection system, the antecedent conditions, the system size, and other factors.

Each facility is different, and each event is different. Because of flow attenuation, travel times, and geographic distribution of storm events, large facilities with large service areas tend to experience less relative flow variations than small facilities. But practically all facilities will still have a large range to handle. For example, some facilities may experience peak flows of three, four, or five times the average flow, and some can be much higher than this.

The selection of a facility's peak design flow is an influential decision, and it affects energy use and power. The peak flow selection may reflect many considerations, including costs, risk tolerances, codes, and/or consent agreements. Water resource recovery facility pumping systems often handle their flow variations in one of the following ways:

- Use of constant-speed pumps each sized for the peak flow, with on–off operation;

- Use of multiple sizes of constant-speed pumps;
- Use of VFDs to operate the pumps at various speeds;
- Use of storage or equalization facilities; or
- Some combination of the above.

4.0 SYSTEM HEAD CURVES AND PUMP PERFORMANCE CURVES

4.1 Overview

To understand pumping power and energy, it is important to have a clear understanding of head, system head curves, and pump performance curves (see Figures 5.2 and 5.3). A system head curve represents the characteristics of the piping and reservoir system (not the pump); it shows the amount of differential head that the system will *require,* if a particular flow is to be achieved. A pump performance curve represents the characteristics of a particular pump; it shows the amount of differential head that that pump will *provide,* for any particular flow.

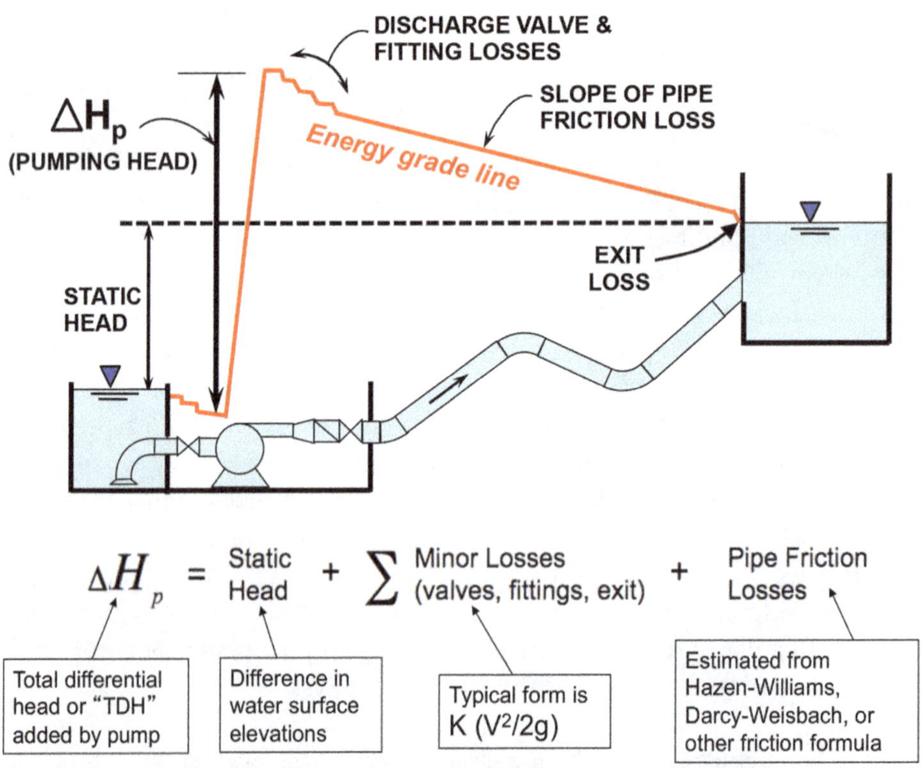

FIGURE 5.2. Pumping system schematic

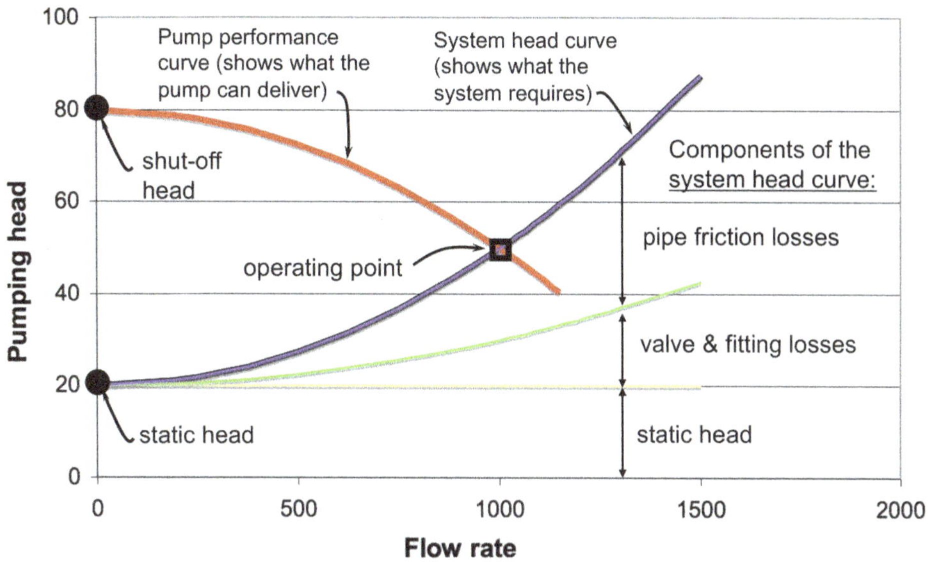

FIGURE 5.3. System head curve and pump performance curve

The actual operating point is where the two curves intersect—it can be viewed as the point at which the pump and the system *agree* to operate together. Note that both the pumping head and the system head, as plotted on these curves, are *differential* heads. Each can be viewed as the *difference* in the total head across the pump, from its immediate suction side to its immediate discharge side. As shown in Figure 5.2, this difference in head is also the amount by which the pump boosts the energy grade line (EGL) of the system, i.e.,

$$\Delta H_p = EGL_2 - EGL_1 = \left[z_2 + \frac{p_2}{\rho g} + \frac{v_2^{\,2}}{2g}\right] - \left[z_1 + \frac{p_1}{\rho g} + \frac{v_1^2}{2g}\right] \qquad (5.8)$$

Where

ΔH_p = pumping head (units of length),
EGL = energy grade line (units of elevation),
Point 2 represents the immediate discharge side of pump,
Point 1 represents the immediate suction side of pump,
z = elevation of the pipe centerline (units of length),
p = pressure of the fluid (units of force/area),
ρ = density of the fluid (units of mass/volume),
V = mean velocity of the fluid (units of length/time) = Q/Area, and
g = acceleration as a result of gravity (units of length/time^2).

4.2 System Head Curves

The system head curve shows the differential head that the system requires of a pump. This curve can be computed as the sum of three components: the static head, the pipe friction losses, and the minor losses (see Figures 5.2 and 5.3). Any of these three components can be a dominant or a negligible part of the total system head (depending on individual system layout and geometry). Accordingly, any of these three can be responsible for a large part or a negligible part of the total energy use and input power. Specific opportunities to reduce this energy use and power are discussed later in this chapter.

4.2.1 *Static Head*

The static head is the component of total system head that exists when the flow from the pump is zero. For simple systems, this is the difference in elevation between the water surface at the starting tank or reservoir (e.g., a wetwell) and the water surface at the destination (e.g., a high point manhole, outfall, or treatment process tank). In many facilities, there is a range of static heads because the wetwell surface can rise and fall as pumps turn off and on, and because the receiving reservoir surface elevations can also fluctuate. Also, in compound or branching forcemain systems, the static head can be influenced by the operation of pumps at other locations.

4.2.2 *Pipe Friction Losses*

The pipe friction losses within a forcemain or piping system result from the interaction of the flowing fluid and the pipe wall. Pipe friction losses are commonly estimated using an empirical formula such as the Hazen-Williams, Manning, or Darcy-Weisbach equations. Detailed discussions of those formulas can be found in many engineering texts. The Darcy-Weisbach formula is shown below (where f = empirical friction factor, L = pipe length, D = pipe inside diameter, v = flow velocity, and g = gravity),

$$\text{Pipe friction loss} = f(L/D)\ (v^2/2g) \tag{5.9}$$

Friction losses are quite sensitive to the flow velocity in the pipe; the calculated losses will vary approximately with $v^{1.85}$ to v^2, depending on the empirical formula used.

For long forcemains, and forcemains with relatively high velocities, the friction losses can be the dominant component of the total pumping head (and can thus account for the majority of the pumping energy use and input power). For short forcemains or low-velocity forcemains, the friction loss may be quite small, or even insignificant. See the next sections of this chapter for opportunities to reduce friction-related energy use and power.

4.2.3 Minor Losses

The minor losses (valve & fitting losses) occur at non-uniformities in the piping system, such as at check valves, isolation valves, bends, tees, reducers, increasers, and exits to reservoirs. Minor losses are proportional to the square of the flow velocity, and are commonly estimated as follows, where K is a loss coefficient applicable to a particular type of valve.

$$\text{Minor loss} = K\,v^2/2g \tag{5.10}$$

Most of the minor losses are typically located in or near the pumping station itself—this is because many valves and fittings are situated near the pumps and header piping. Also, these valves and fittings (like the pumps themselves) typically have smaller diameters and, therefore, higher velocities, than those in the main discharge pipe or forcemain. These smaller valves and fittings thus create significantly higher losses.

Check valves, because of their particularly high loss coefficients, can contribute a large portion of the minor losses. Even larger energy losses can occur if isolation valves are throttled to a part-closed position to force a constant-speed pump to produce a lower flow. This practice should be avoided when possible because it imposes an artificially high system head onto the pumps and can thus significantly increase energy use.

4.3 Pump Performance Curves

Figure 5.4 is an example of a pump performance curve from a manufacturer's catalog. This example shows a pump at a constant operating speed (1185 rpm) and shows a range of different impeller diameters that are available (in this case, ranging from 14.50 in. to 17.50 in.). A pump performance curve also typically shows the mechanical efficiency of the pump (e_p). For the pump in Figure 5.4, the efficiency ranges from approximately 49% to 80%.

Figure 5.5 is another example of a pump performance curve. This curve is a variable speed curve; it shows one impeller size at a range of different speeds (in this case, from 593 rpm to 1185 rpm). As exemplified in Figure 5.5, a pump's performance curve is greatly affected by its operating speed. If the speed of a given pump is increased or decreased, the new performance curve can be estimated as follows (where Q = flowrate, H = pumping head, P = power, and n = rotational speed). These are commonly known as the pump affinity laws.

$$\frac{Q_2}{Q_1} = \left[\frac{n_2}{n_1}\right] \qquad \frac{H_2}{H_1} = \left[\frac{n_2}{n_1}\right]^2 \qquad \frac{P_2}{P_1} = \left[\frac{n_2}{n_1}\right]^3 \tag{5.11}$$

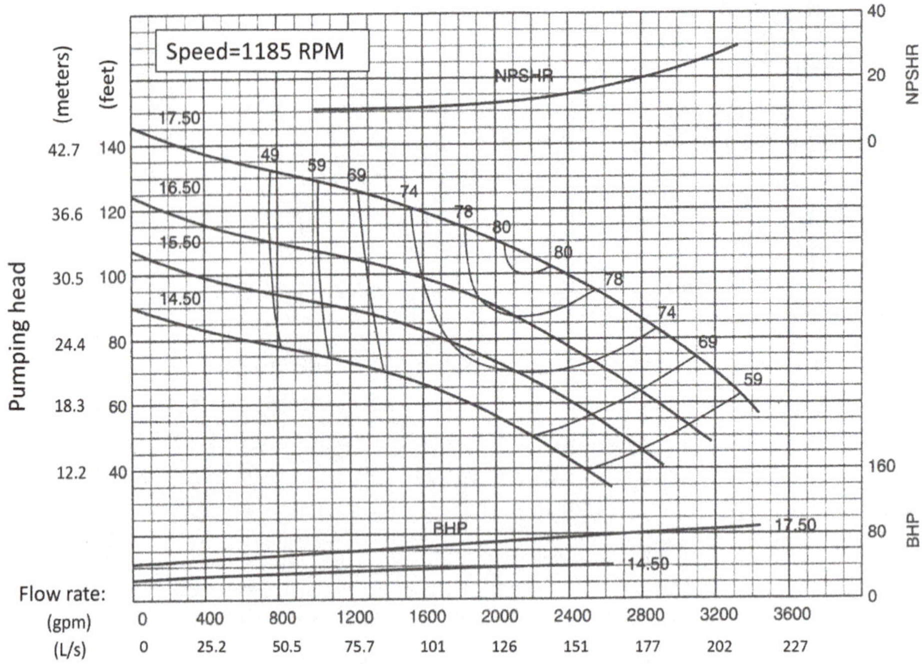

FIGURE 5.4. Example of a pump performance curve (constant speed, with different available impeller sizes)

Although the affinity laws allow pump performance curves to be projected for a variety of different speeds, it must be remembered that the actual operating point will still be greatly dependent on the system curve, and will not be simply proportional to speed. For example, in Figure 5.5, consider the effect of reducing from 100% speed (1185 rpm) to 80% speed (948 rpm). With System Curve A, the 80% speed would reduce the pumped flow from 600 gpm (37.85 L/s) to 380 gpm (23.97 L/s)—approximately 63% of the full speed flow. With System Curve B, the 80% speed would reduce the pumped flow all the way to zero, because the pump curve has dropped below the system curve.

4.4 Equipment Efficiencies—Pump, Motor, and Variable Frequency Drive

4.4.1 Pump Efficiencies

For any centrifugal pump running at a particular speed, there is an optimum operating point of maximum efficiency. This is commonly called the BEP, or best efficiency point. For example, in Figure 5.4 the BEP is slightly higher than 80%; in Figure 5.5 the BEP is slightly higher than 73%. Operating a

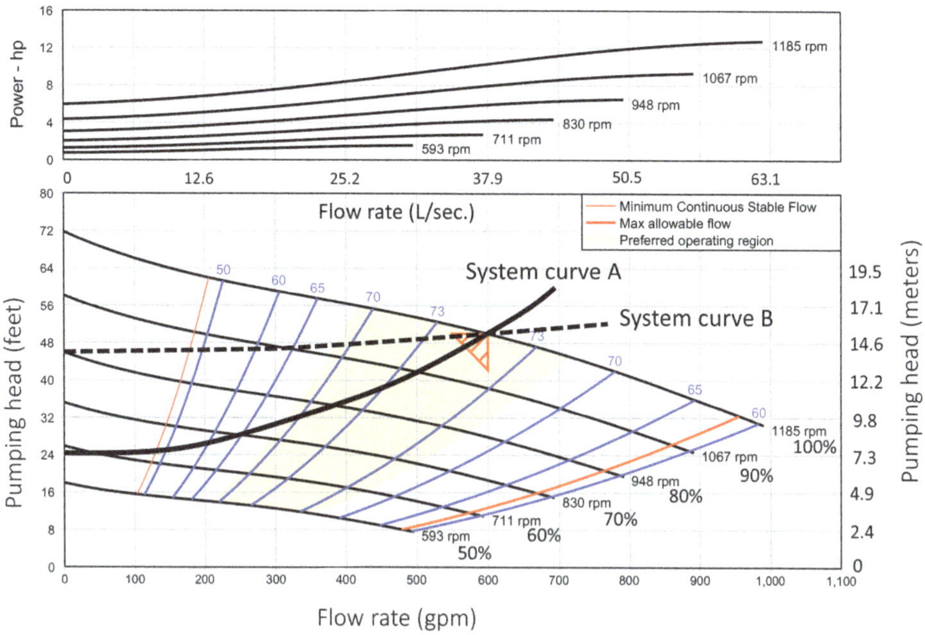

FIGURE 5.5. Example of a variable-speed pump performance curve, with one selected impeller

pump at a high mechanical efficiency, close to the BEP if possible, can help to reduce power and energy (assuming the same flow and head), as shown in Equations 5.2 through 5.7. Also, the higher efficiency will yield a smoother mechanical operation and longer operating life for the equipment.

For various reasons, however, pumps can often be found operating far away from their BEPs. Sometimes a pump selection, intending to be conservative, is sized for a future assumed condition in which the pipe walls are assumed to have become much rougher. For the initial operation, however (and perhaps for many decades), the roughness and friction may be much lower. This may cause the pump to operate far to the right of its BEP (at a higher flow, at a lower efficiency, and drawing higher input power, than planned). In other cases, the wearing of a pump impeller, partial clogging of a pump, valve, or pipeline, air pockets in the piping, cavitation damage, or alignment problems may detract from a pump's performance and efficiency. In still other cases, it may be found that the original pump, or its impeller, has been replaced with one of differing characteristics.

There is a wide range of efficiencies in the universe of centrifugal wastewater pumps—most are between 60% and 90%, and some are between 70% and 80%. Large flow pumps are often able to offer higher BEP efficiencies than are smaller flow pumps. This is because the large pumps have larger passageways and less internal losses. Wastewater pumps must be engineered to handle a wide range of solid materials in the flow, and are typically rated to pass at least a 3-in.-diameter sphere. This constraint requires special impeller designs, and often some compromise in pump efficiency, compared to a clean water pump of the same size.

4.4.2 Motor Efficiencies

The efficiency of the electric motor is also important to energy use and power draw (see Equations 5.2 through 5.7). In wastewater pumping systems, electric motor efficiencies are typically higher than (and fall within a tighter range than) the pump efficiencies. For example, when operating at full load, electric motor efficiencies can often be found in the range of 90% to 95%. It should be noted however that electric motor efficiencies can decline when they are operating at reduced speeds and reduced torques, such as with VFDs. See Chapter 3 for a more detailed discussion of electric motors and their efficiencies.

4.4.3 Variable Frequency Drive Efficiencies

In addition to the pump and the motor, the efficiency of the VFD, if used, affects energy use and power draw for a pumping system. When operating at full speed, it is common for the efficiency of the VFD to be high—such as 97% to 98%. When VFDs are operating at reduced speeds and loads, this efficiency may decrease. The amount by which the efficiency decreases should be verified with the VFD manufacturer. See Chapter 3 for a more detailed discussion of VFDs and their efficiencies.

5.0 OPPORTUNITIES TO REDUCE PUMPING ENERGY USE AND POWER DEMAND

Pumping energy use and input power depend on the four key variables discussed earlier: flow, pumping head, equipment efficiencies, and run times. Therefore, to reduce pump energy use and/or input power, one should look for opportunities to either reduce flow, reduce pumping head, increase equipment efficiencies, decrease run times—or some combination thereof. Several such opportunities are discussed below.

When evaluating possible energy and power savings, it is useful to recall that there are different types of system curves. Some systems are dominated by static head, with little friction loss (e.g., consider a short discharge pipe/forcemain and a large difference in reservoir elevations). Other systems are dominated by friction losses and minor losses, with little static head (consider a long forcemain, with only a small difference in elevation). When a system is primarily composed of static head, there will generally be less opportunity to save energy (kWh) because there is not much friction loss to be saved. Because each actual pumping system is different, calculations should be done for each set of conditions to determine potential savings.

5.1 Operating Elevations

In some cases (but not in others), there may be some room to save energy and power by increasing the operating levels of the wetwell (and/or decreasing those at the receiving chamber). This will lower the static head and decrease the energy use and power required. Of course, such adjustments should only be done if they will not adversely affect the performance of the facilities or increase the risks of a sanitary sewer overflow.

5.2 Forcemain and Piping Design Considerations

In general, the selection of a larger pipe diameter during design (for forcemains, valves, and fittings) reduces the velocity in those pipes (for the same flow), and thus reduces the headlosses, energy use, and input power (see Sections 4.2.2 and 4.2.3). Of course, the larger pipe sizes also increase the capital cost of the piping and valves.

Particularly for longer forcemains, selecting an optimum pipe diameter requires balancing two opposing constraints. On the one hand, the pipe size needs to be large enough for peak flows, thus keeping the peak velocities, friction losses, pumping heads, motor sizes, and energy costs from becoming excessively large. On the other hand, during normal flow situations (which exist most of the time), a smaller forcemain diameter makes it easier to maintain velocities higher than 0.61 to 0.91 m/s (2 to 3 ft/s), which helps to keep solids in suspension.

Sometimes, several alternative pipe sizes may be reasonable choices for a forcemain and for valves/fittings. To help make this decision, the energy and power costs associated with the different pipe sizes, together with the different capital costs, should be incorporated into an economic analysis comparing the alternatives. A life-cycle analysis should be performed so that energy and maintenance costs are part of the evaluation.

The roughness of the interior pipe wall also affects friction losses and energy use. Some pipe materials, such as plastics, are smoother than others, and would be expected to create lower friction losses, assuming the same flow and diameter. When comparing pipe materials, however, the designer should also note the actual inside diameters because they can be different from one another. For example, a nominal 30 cm (12-in.) PVC pipe may have a smaller inside diameter than a nominal 30 cm (12-in.) ductile iron pipe, and this will affect losses and energy use.

5.3 Pump Selection—Size, Number, Speed, and Operation

Energy saving opportunities are affected by the selection of the size, number, speeds, and operation of the pumps to be included in a pumping facility. Several different pumping arrangements are discussed below. In all cases, individual pumps should be selected so that the expected operating range of the pump is as close as practical to the best efficiency point.

5.3.1 Two Equal Constant-Speed Pumps

This is the most common arrangement found in many small pumping facilities. With only two pumps, each pump is normally sized for the peak flow, because of the need for standby capacity. Codes typically require that a pumping facility must be able to achieve its peak design capacity even if one pump is out of service.

A two-pump constant-speed facility will typically cycle on and off many times throughout the day (sometimes as many as 50 or more cycles per day). Many smaller facilities operate adequately in this way. A drawback of this arrangement, however, is that every drop of wastewater will be pumped against the maximum pumping head of a peak flowrate, whether needed or not. If the discharge forcemain is long, and friction loss dominates over static lift, this will translate into high energy use and high input power. Also, downstream facilities or processes will receive the flow in intermittent blocks of artificially high flowrates. In some cases, this may be acceptable (e.g., a collection system pumping station discharging to a gravity manhole), but in other cases it may not (e.g., for some biological or physical treatment process within a WRRF).

5.3.2 Three or More Constant-Speed Pumps, of Various Capacities

Larger pumping facilities often include three or more pumps, to provide additional flexibility. In some cases, this can also provide energy and power savings, because some of the pumps can be sized for intermediate pump capacities (rather than peak flow). The small or intermediate pumps will typically receive the most run time, and their lower flowrates can (in some cases) result in lower pumping heads and thus lower energy usage and lower input power. They will also cycle on and off less frequently than a two-pump constant-speed alternative.

The use of multiple intermediate pump capacities can thus provide many of the same benefits as the use of VFDs, as discussed in the next section (although the pumping flow range will occur in multiple steps rather than over a continuous range).

5.3.3 Variable Speed Pumping with Variable Frequency Drives

Variable frequency drives modulate the frequency of the electric power supplied to the motor. This allows the motor and pump to operate through a range of speeds, pumping only as much flow as needed at the time. An example of a variable speed pump curve is show in Figure 5.5.

For many facilities, the use of VFDs is an effective way to achieve flow flexibility. In some cases (when friction losses are relatively large), the use of VFDs can provide for lower energy use and input power (compared to a two-pump constant-speed operation). In other cases, it will not.

The system head curve has a large influence on whether or not a VFD-equipped system can save energy. For system curves in which pipe friction losses and minor losses at peak flow are relatively high (i.e., steep system curves, long forcemains, high peak velocities), the VFD system can provide considerable savings (relative to a peak-flow constant-speed system). This is because the VFD allows the pumps to run at reduced speeds, producing only as much flow and head as needed, and thus avoids (most of the time) the high pumping heads of peak flows.

On the other hand, for system curves in which the friction losses/minor losses are small (i.e., high static heads, flat system curves, short forcemains, low peak velocities), there is little opportunity to save energy (kWh) with a VFD. In this case, the pumping head remains nearly unchanged at the different flowrates. The reduced flowrates are offset by longer run times, and the energy use per day remains nearly the same (see Equation 5.5). It is worth noting, however, that the input power (kW) *is* still reduced in this case, because power is proportional to flowrate and does not depend on run time (see Equation 5.4). Therefore, the VFD may provide savings in the peak kW demand charges even when the total energy use (kWh) is not reduced.

When evaluating energy use and input power at reduced speeds, it is important to observe the actual changes that occur in the key parameters (flow, pumping head, and equipment efficiencies). Each set of system curves and pump curves is different (see Figure 5.5, for example). In some cases, the efficiencies of the pump at reduced speeds can be significantly different than at the full speed as a result of the intersection with the system curve. Also, when speed and load are reduced, the motor efficiency and the VFD efficiency can decrease.

Note that the decision to use VFDs generally does not affect the required sizing of the peak flow pump capacity. The same peak flow pumps would typically be selected for a VFD-driven system as for a constant-speed system.

5.3.4 Parallel Pumping

Using multiple pumps together in parallel to meet peak flows can be a useful feature, particularly for larger flow facilities with three or more pumps. Parallel pumping can be used for either constant-speed or variable-speed systems. For parallel pumping to be effective, pumps must be specifically selected for this purpose. A given pump may operate at different flows, pumping heads, and efficiencies when used individually versus in combination with multiple pumps. The designer may thus need to decide whether to have the best efficiencies achieved when a single pump is running, or when multiple pumps are running.

Systems with steep system curves (such as long forcemains and/or high peak velocities), are not typically a good match for parallel pumping. In such cases, each of the pumps (when operating together) will be pushed far back onto their pump curves, and each will operate at a greatly increased head and greatly reduced flow. In some cases, two equal pumps together would produce only slightly greater combined flow than one pump alone.

5.4 Use of Storage or Equalization Facilities

The use of storage or equalization facilities upstream of a pumping facility can affect its energy use (although energy savings is not their primary purpose). Although almost all pumping systems have some storage in the wetwell or sewer system, it is typically not large enough to reduce the peak pumping capacity. If the storage volume is made large enough, however, it can reduce the peak pumping capacity needed for high flow events.

From a mechanical perspective, storage provides a simple, reliable method of managing variable flows—because much of the variability is removed from the pumps and taken up by the storage reservoir. The pumps can then be selected for longer run times at more moderate flows and heads (without the large spikes of short-term peak flow), and the total energy use and power demands will be reduced. Of course, large storage or equalization facilities also have disadvantages, such as large capital costs, land requirements, odor potential, and the need to deal with solids. As with any capital project, the estimated energy costs and operating costs should be included, along with capital costs, in an economic analysis.

5.5 Reducing Inflow and Infiltration

Efforts to reduce the inflow and infiltration (I/I) of clearwater into a collection system is an ongoing, important activity for many systems. The degree to which a system is subject to stormwater inflow and groundwater infiltration can have a great effect on the design, capital cost, and sizing of sewers, pumping stations, and treatment facilities. Successful reduction of I/I has numerous benefits, including reduction of sanitary sewer overflows, reduction of capital costs, and reduction of pumping energy use and power demand.

The infiltration of groundwater into sanitary sewers, even during dry weather, can be quite substantial in many systems. In some cases, this can be 10%, 20%, 30%, or even more of the total average daily flow entering a WRRF. Stormwater inflow, as previously discussed, can be much larger still (although for shorter periods), and is responsible for the extremely high peak flows (and the high peak energy use) that are experienced at many facilities.

Successful reduction of I/I is not easy, and is difficult to quantify, because of the large areas covered by collection systems, and because each wet weather event is different in intensity, distribution, and antecedent conditions. However, when successfully accomplished, it is effective in reducing pumping energy use. The reduction of I/I is different than the other energy saving alternatives discussed previously in this chapter, because it reduces the actual annual flow volumes and the peak flows that are received into a facility.

5.6 Water Conservation

The average water use per person in the United States has been decreasing in recent decades. The declining trend has been evident in practically all states, and is the result of a variety of reasons, including the increased presence of reduced flow toilets and fixtures, local and state water conservation regulations, public education/awareness, manufacturing trends, and increasing prices for water. A detailed discussion is beyond the scope of this text, but this trend is likely to continue, and it has had a helpful effect on pumping energy use, because it has lowered the total flow volume that would otherwise have been received (for the same population).

6.0 EXAMPLE CALCULATIONS

Consider a wastewater pumping station with an average daily flow (ADF) of 37.85 L/s (600 gpm, 864 000 gal/d). Its peak pumping capacity is 151.4 L/s (2400 gpm, 3.456 mgd). The station has two large pumps (each rated at 151.4 L/s or 2400 gpm) and one small pump (rated at 75.7 L/s or 1200 gpm). Pump curves and system curves are shown in Figure 5.6.

The pumps operate at constant speeds, and only one pump operates at a time. Currently, the small pump does practically all the pumping. It cycles on and off, and its total run time is 12.0 hours on an average day. The large pump runs only infrequently, during wet weather events.

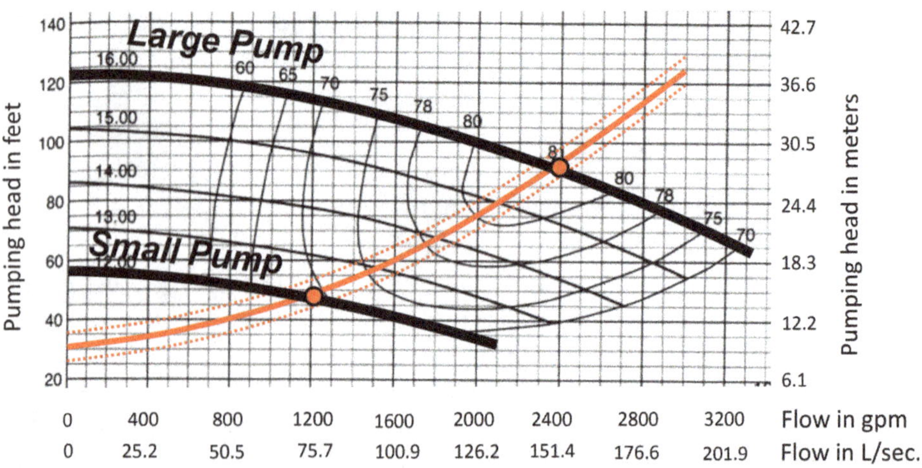

FIGURE 5.6. Example problem

The owner is considering whether it would be more energy efficient to run the large pump instead of the small pump, on a regular daily basis. It is noted that the large pump has a much higher efficiency (0.81) than the small pump (0.69). Also, the large lump would only need to run 6.0 hours per day to meet the ADF (rather than 12.0 hours with the small pump). For this example, assume that the electric motors have an efficiency of 0.95 for both pumps.

To evaluate the above, apply Equations 5.6 and 5.7, as follows.

If the large lump handles the entire daily flow:

- Operating condition = 2400 gpm @ 28 m (91 ft.)
- Pump efficiency e_p = 0.81. Motor efficiency e_m = .95
- No VFD in this example, therefore e_{vfd} = 1.00
- Pump run-time per day = 864 000 gal/2400 gpm = 360 minutes = 6.0 hours
- Input power = (.000189)(2400 gpm)(91') / (.81)(.95)(1.0) = 53.64 kW
- Energy used per day = (53.64 kW)(6.0 h) = 322 kWh

If the small pump handles the entire daily flow:

- Operating condition = 1200 gpm @ 15 m (48 ft)
- Pump efficiency e_p = 0.69. Motor efficiency e_m = .95
- No VFD in this example, therefore e_{vfd} = 1.00
- Pump run time per day = 3 270 596 L (864 000 gal) / 1200 gpm = 720 minutes = 12.0 hours

- Input power = (.000189)(1200 gpm)(48') / (.69)(.95)(1.0) = 16.61 kW
- Energy used per day = (16.61 kW)(12.0 h) = 199 kWh

Conclusion: despite the fact that the small pump has a considerably poorer efficiency, and requires twice the run time, it will still use much less energy than the large pump to handle a full day of flow (199 kWh vs. 322 kWh). This is because the small pump operates at a much lower pumping head than the large lump (48′ vs. 91′). Energy use depends on the *combined* effects of Q, pumping head, efficiency, and run time. In this example, the lower pumping head of the small pump more than compensates for its poorer efficiency, and its longer run time is simply offset by its lower flowrate. Also, the small pump draws much less input power than the large pump (16.61 kW vs. 53.64 kW). The lower energy use and the lower power demand will likely both be reflected in the electric bills for this station.

7.0 OPERATION AND ENERGY MANAGEMENT PRACTICES FOR PUMPING

Regular tracking of the total energy use and the monthly electric bills for a pumping facility is a valuable practice. In particular, the energy use per volume pumped is a helpful indicator to identify problems that may be developing.

For various reasons, it is possible for a pumping system to use much more energy than needed (e.g., poor matching of a pump with its system curve, pump damage, wear and tear, partial pump clogging, valve clogging, forcemain blockages, air pockets, and other causes). To verify the point at which the pump is operating, it is necessary to measure the actual flowrate and the actual pumping head produced by the pump. Flow can be measured by a flowmeter if one exists, or by a timed wetwell drawdown volume test. Pumping head can be estimated by measuring the pressure, elevations, and computed velocities, at the suction and discharge side of the pump. To determine the actual efficiency of the pumping system, the flow of electrical power into the system must also be measured. The results can then be compared with the expected system head curves and pump performance curves, and problems can be diagnosed.

8.0 SLUDGE PUMPING ENERGY CONSIDERATIONS

The pumping of thickened sludges can be significantly more difficult than for raw wastewater, mixed liquor, or return activated sludges. Because thick sludges are not true fluids, their flow and viscosity behavior can be non-continuous, and their friction losses can be much greater than for water, wastewater, or thin sludges. Energy use and input power still depend on flow, pumping head, equipment efficiencies, and time, as arranged in Equations 5.2 through 5.7. However, the pump curves, system curves, and equipment efficiencies can be different, because thickened sludge pumping often requires the use of positive displacement or other pump types, rather than centrifugal pumps. Positive displacement pumps often have steep performance curves, in which flowrate remains nearly constant across a wide range of heads. This is needed because of the complex and sometimes uncertain nature of sludge pumping heads. Higher equipment costs, higher maintenance costs, higher heads, and lower efficiencies are commonly associated with thickened sludge pumping. For a detailed discussion of thickened sludge pumping, see the suggested readings at the end of this chapter.

9.0 CASE STUDY: UPGRADE TO PUMPING STATION NO. 15, MADISON METROPOLITAN SEWERAGE DISTRICT

Pumping Station No. 15 (PS15) is one of the 18 stations in the Madison Metropolitan Sewerage District's wastewater collection system. PS15 was originally constructed in 1974, and it received a comprehensive upgrade in 2017. The upgraded station includes new pumping equipment; new piping and valves; new heating, ventilating, and air conditioning (HVAC) systems; new electrical and control systems; and other improvements. The upgrade allows the station to be well positioned for long-term reliability, flexibility, energy efficiency, and future growth.

The project also reflects considerable collaboration with the local community and businesses. Located adjacent to a popular park with access to Lake Mendota, the upgraded PS15 incorporates improved site aesthetics, public restrooms, drinking fountain, bike racks, bike repair stand, bioretention basins, and an invasive species boat wash station. The PS15 upgrade project was the winner of the first Envision Gold Award for Sustainable Infrastructure in the State of Wisconsin. Photos of the project are shown in Figures 5.7 and 5.8.

The new pumping system at PS15 includes three equal VFD-driven pumping units, together with new piping, valves, and controls. Each pumping unit includes an 8-in. horizontal dry-well pump with 125-HP, 1190 rpm motor. The approximate pumping system curves are shown in Figure 5.9.

FIGURE 5.7. New exterior of Pumping Station No. 15. Reprinted with permission from Madison Metropolitan Sewerage District.

FIGURE 5.8. New pumping equipment at Pumping Station No. 15. Reprinted with permission from Madison Metropolitan Sewerage District.

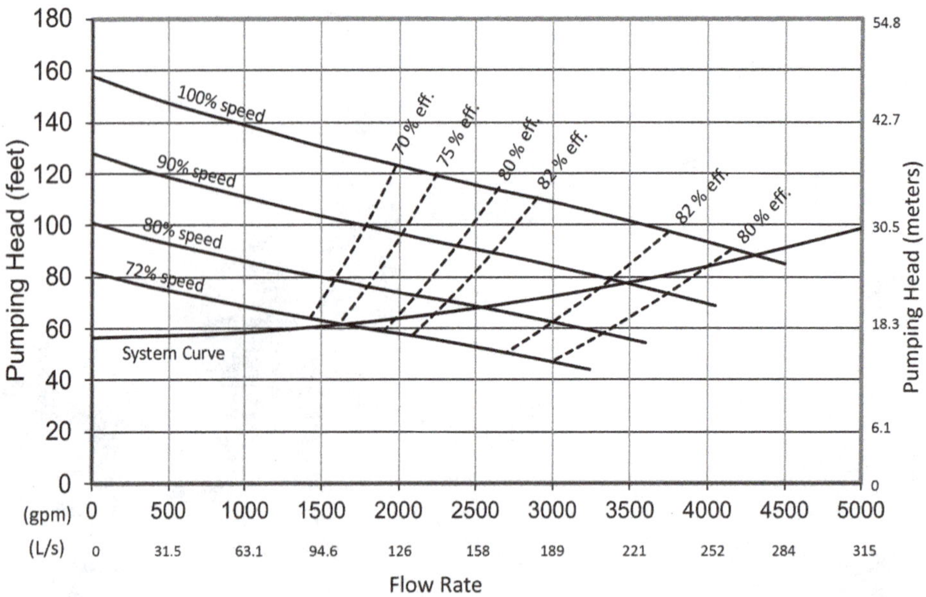

FIGURE 5.9. Approximate pump and system curves for Pumping Station No. 15

Each pump can operate through a range of speeds, from approximately 72% to 100% of full speed. One pump normally operates at a time, and the operating pump is alternated with each run cycle.

Key PS15 energy-related data for calendar year 2019 are summarized below.

- Total 2019 pumped volume = 2031 ML (536.61 mil. gal)
- Average daily flow = 5.56 ML/d (1.47 mgd)
- Total 2019 pump run time = 5723 hours
- Total 2019 facility metered energy use = 225 851 kWh
  - Estimated energy used for pumping = 174 000 kWh
  - Estimated energy used for HVAC and ancillary equipment = 52 000 kWh
- Total 2019 energy use per volume pumped = 111 kWh/ML (421 kWh/mil. gal)
- Total 2019 electric utility charges = $36,387
  - kWh charges = $14,246
  - Demand charges = $17,434
  - Base charges = $4,706

Under normal dry-weather conditions, the pumps are set to operate at the low 72% speed, providing a discharge of approximately 100 L/s (1600 gpm,

2.3 mgd). This is more than adequate for the station's average daily flow. Under normal conditions, therefore, the pumps turn on and off throughout the day, allowing the wetwell to fill and draw down, similar to a constant-speed station. In addition, the pumps are programmed to run for one cycle each day at full 100% speed, as a "flushing cycle." The higher velocities of the flushing cycle help to minimize potential solids accumulation in the forcemain and wetwell.

During high flow wet weather events, any individual pump can increase its speed as needed to match the incoming flow, up to approximately 270 L/s (4300 gpm, 6.2 mgd). In addition, the pumping system is designed to handle expected future growth in the service area. If needed, multiple pumps can be operated together in parallel to achieve a peak discharge up to approximately 370 L/s (5800 gpm, 8.3 mgd).

The use of the VFDs allows PS15 to generally pump at a flowrate that is only as high as needed. This provides a significant energy advantage for PS15. The vast majority of the flow volume can be handled at the low 72% speed, which produces approximately 1600 gpm at a relatively low head of approximately 62 ft (see Figure 5.9). At this flow, the headlosses in the system are small, and the pumping head is only a few feet higher than the static head. By contrast, the pumping head at full 100% speed would be nearly 90 ft. As discussed earlier in this chapter (see Equation 5.7), energy use is proportional to flow, head, and elapsed time, and inversely proportional to equipment efficiencies. Although the equipment efficiencies (pump, motor, and VFD) do tend to be lower at the 72% speed than at the 100% speed, this is more than offset by the large savings in pumping head in this case.

The upgraded PS15 project also provides improved capabilities to monitor a variety of parameters associated with energy management, including instantaneous power draws, metered flowrates, gage pressures, individual pump run times, and other data. Metrics such as the energy use per volume pumped (kWh/MG) provide a useful snapshot of the energy intensity for the facility. The data from the initial years of operation will provide a valuable baseline for present and future energy management at PS15.

10.0 SUGGESTED READING

European Commission. (2001). *Study on improving the energy efficiency of pumps.*

Hydratek, Ontario Power Authority. (2013). A Pump Efficiency Assessment and Awareness Pilot Study.

Hydraulic Institute. (2004). *Variable speed pumping—A guide to successful applications.*

Karassik, I. (2007). *Pump handbook* (4th ed.). McGraw-Hill Education.

Paschke, N. W. (2012). Performance under pressure: Practical perspectives on pumping and energy. *Water Environment Federation*, *WET Magazine*, July 2012.

Paschke, N. W. (2018). Analysis of trends in municipal per capita water use. *Singapore International Water Conference*, July 2018.

Sanks, R. L. (Ed. in Chief). (1998). *Pumping station design*. Butterworth-Heinemann Publishers.

Tchobanoglous, G. (1981). *Wastewater engineering: Collection and pumping of wastewater*. McGraw-Hill.

Volk, M. W. (2005). *Pump characteristics and applications*. CRC Taylor & Francis.

Water Environment Federation. (2009). *Energy conservation in water and wastewater facilities* (WEF Manual of Practice No. 32). WEF Press, McGraw-Hill.

Water Environment Federation. (2017). *Operation of water resource recovery facilities* (6th ed.; WEF Manual of Practice No. 11). WEF Press, McGraw-Hill.

Water Environment Federation. (2018). *Design of water resource recovery facilities* (6th ed.; WEF Manual of Practice No. 8). WEF Press, McGraw-Hill.

6

Mixing

Ed Kobylinski; Peter V. Cavagnaro, PE, BCEE, PMP; James Fischer; & Ethan Willats

1.0 INTRODUCTION

Information for mixing was developed as a separate chapter, rather than repeat information for every process or operation that uses energy. Mixing roles at a water resource recovery facility (WRRF) have changed with the addition of biological nutrient removal (BNR) processes. With 15% to 40% of the total activated sludge basin volume dedicated to anaerobic and anoxic zones for nutrient removal, mixing energy becomes a significant energy demand at BNR facilities.

The design for activated sludge mixing has always been a power per unit volume assumption. Older designs from the 1980s to the 1990s, did not take into account the different efficiencies of mixing supplied by different styles of mixers. Consequently, many mixing applications were significantly overpowered. The discussions in this chapter will briefly touch on the theory of mixing to give the reader an appreciation for mixing relationships and that not all nameplate horsepower mixers are equal. Some are more efficient than others. Another takeaway for the reader is how necessary it is to first define the level of mixing desired for the specific application and that communication with the mixer supplier is crucial. Mixer placement and mixer orientation is critical for good efficient mixing. Other mixing applications like mixing sludge holding tanks require coordination with the mixer manufacturer as tank shape and liquid level variations affect mixer selection and energy input. Computational fluid dynamic modeling may be needed for mixer sizing and to verify mixer placement to get the desired mixing.

2.0 MIXING APPLICATIONS AND ENERGY INPUT

Mixing is used at a WRRF to disperse chemicals and maintain solids in suspension. Some of the applications in which mixing is used are listed in Tables 6.1 to 6.4.

3.0 FACTORS AFFECTING POWER REQUIRED FOR MIXING

As illustrated in Tables 6.1 to 6.4, there are many possible applications for mixing in a wastewater treatment process. The amount of energy required depends on several factors, including: the process requirement, the type of

TABLE 6.1. Common mixing applications at a WRRF: preliminary treatment

Process or Operation	Type of Mixing	Purpose	Observation
Raw wastewater lift station wetwell	Diffused air Recirculation of flow	• Maintain solids and flotables in suspension • Disperse chlorine used for pre-treatment	Not very common
Equalization basins	Diffused air	• Maintain solids in suspension • Maintain freshness of wastewater	
Preliminary treatment	Diffused air	• Maintain organic solids in suspension	

TABLE 6.2. Common mixing applications at a WRRF: primary and secondary treatment

Process or Operation	Type of Mixing	Purpose	Observation
Primary treatment	Air sparger	• Mixing of scum well	—
Biological treatment —diffused air activated sludge	Top entry mixer Submersible mixer Large bubble mixer	• Anoxic zone • Anerobic zone	
	Diffused air Large bubble mixer Submersible mixer	• Mixing MLSS	
Biological treatment —high purity oxygen	Top entry mixer	• Mixing MLSS	
Biological treatment —oxidation ditch	Horizontal brush style aerators	• Mechanical aeration • Impart velocity to MLSS to travel around ditch • Maintain MLSS in suspension	Speed must be maintained high enough to keep MLSS from settling
Process or Operation	**Type of Mixing**	**Purpose**	**Observation**
Channel aeration	Diffused air	• Maintain MLSS in suspension	
Phosphorus removal (alum or ferric chloride feed)	Hydraulic Mechanical—top entry mixer	• Rapid mixing of chemical for precipitation of phosphorus	
	Hydraulic	• Flocculation of phosphorus precipitate	
Secondary clarifier	Top entry mixer	• Flocculation of MLSS in stilling well	

mixing used, the size of the vessel in which mixing is required, specific gravity of the solids, and the viscosity of the solution being mixed.

3.1 Understanding Process Requirements and Mixing Level

The first step in energy optimization of a mixing application is selection of the correct level of mixing intensity for the application.

- Rapid mixing for chemical addition requires shear applied to the liquid stream;

TABLE 6.3. Common mixing applications at a WRRF: tertiary treatment and disinfection

Process or Operation	Type of Mixing	Purpose	Observation
Sand filtration	Top entry mixer	• Disperse chemicals for phosphorus removal	
	Hydraulic mixing		
	In pipe mixing	• Addition of chlorine solution to backwash water	
Disinfection	Aspirator	• Mixing of chlorine gas or liquid with facility water	
	Top entry	• Addition of chlorine solution to wastewater for disinfection	Intense mixing desired with minimal backmixing
	Hydraulic		
Dechlorination	Aspirator	• Mixing of chlorine gas or liquid with facility water	
	Top entry	• Addition of sodium solution for dechlorination	Intense mixing desired with minimal backmixing
	Hydraulic		

- Mixing in a sludge tank requires solids to be fully suspended and a uniform solids concentration for feed to a dewatering system and the tank volume often varies;
- Mixing in an aerobic zone involves dispersing oxygen throughout the basin, but in many BNR systems, air mixing the end of the basin may result in over-aeration; and
- Mixing within a biological nutrient removal basin requires solids to be suspended, but for good BNR performance, mixing does not have to disperse solids as thoroughly as in other applications.

Mixing intensity was evaluated in a field test of the anaerobic and anoxic zones of an operating BNR facility at Rogers, Arkansas. A third BNR train

TABLE 6.4. Common mixing applications at a WRRF: solids management

Process or Operation	Type of Mixing	Purpose	Observation
Thickening	Inline static mixer	• Mix polymer with sludge	
	In pipe mixing through a series of 90-degree bends		
Aerobic digestion	Diffused air	• Maintain solids in suspension	
Anaerobic digestion	Recirculated digester gas distributed through a system of lances	• Maintain digesting solids in suspension • Even temperature distribution	
	Recirculated digester gas distributed through a single or multiple gas lifts		
	Digester gas distributed through a large bubble gas mixing device		
	Hydraulic—sludge recirculated at high flow rats		
	Hydraulic—sludge circulated through high velocity nozzles		
	Top entry mixer		
	Top entry mixers mounted in tubes on the side of the digester		
Liquid sludge storage	Hydraulic or pumped mixing	• Maintain solids in suspension	
	Mechanical mixing		

(*continued*)

TABLE 6.4. (*Continued*)

Process or Operation	Type of Mixing	Purpose	Observation
Dewatering	Inline static mixer In pipe mixing through a series of 90-degree bends Proprietary injection rings	• Mix polymer with sludge	
Sidestream treatment	Diffused aeration Mechanical mixer	• Maintain solids in suspension	
Polymer	Mechanical mixing	• Wet and age dry polymer	
	Mechanical mixing Hydraulic mixing	• Blending of polymer with carry water	

was added to the site and the installed top entry mixers had variable speed drives. The study was to determine the appropriate speed for mixer operation in the anaerobic and anoxic zones. The basin had a top entry axial flow, downward pumping mixer. The mixer design speed was 34 revolutions per minute (rpm) and for the study the mixer was run at 28, 16, and 8 (rpm). Samples of the mixed liquor suspended solids (MLSS) were collected at varying depths in the basin (0.9, 1.8, 2.7, and 3.7 m below the water surface level) and at five different locations within the tank (locations selected for sampling were the most likely locations to be poorly mixed). The variability is defined as the largest concentration difference from a single point divided by the average concentration of all the samples. The variability of MLSS concentrations was 3.35% at 28 rpm, 4.6% at 16 rpm, and 5.5% at 8 rpm (Kobylinski, 2014). As the mixing was reduced (lower speed), the mixture was less uniform. However, the uniformity of MLSS concentrations was quite good even at the lowest speed and a 5.5% variability for the slowest speed is within normal total suspended solids analysis error. The real difference between the mixing regimes at the three speeds was the surface appearance of the MLSS. The pictures in Figures 6.1, 6.2, and 6.3 present the surface appearance of three different levels of mixing observed during the field test.

The major difference in mixing is evident at the liquid surface. Figure 6.1 at the highest speed is a uniform chocolate brown. In Figure 6.2 to the right

FIGURE 6.1. MLSS surface appearance while mixing at 28 rpm. Note MLSS uniformity of color and texture.

FIGURE 6.2. MLSS surface appearance while mixing at 8 rpm (left side of wall) and 16 rpm (right side of wall)

of the wall (16 rpm mixer speed), the uniform color of the MLSS is similar to that in Figure 6.1, while the slowest speed (8 rpm) exhibits a surface with a clear layer of water over the MLSS. Ridges and valleys in the MLSS can be seen in Figure 6.2 to the left of the baffle wall and in Figure 6.3. While still fully mixed at 8 rpm, the MLSS solids are not lifted to the surface and a liquid barrier between the atmosphere and MLSS is formed. The surface condition in Figure 6.3 is more appropriate for mixing of an anaerobic and anoxic zone where oxygenation is not desired. Overmixing in a BNR facility wastes energy and may harm process performance. Exposing an MLSS that is supposed to be anoxic or anaerobic to the surface of the basin, exposes the MLSS to oxygen that could interrupt the anaerobic and anoxic processes. The Roger WRRF was meeting their phosphorus limit with this mixing strategy.

A BNR facility's sensitivity to overmixing can be affected by the influent wastewater rapidly biodegradable chemical oxygen demand (rbCOD) concentration. For WRRFs with a high rbCOD concentration, there is enough driving force to remove any dissolved oxygen transferred to the MLSS and process performance will not be affected significantly. However, as more influent wastewaters are being characterized for rbCOD, many WRRFs are finding that for a good portion of the year, if not for the whole year,

FIGURE 6.3. Close-up of MLSS surface appearance with mixing at 8 rpm. Basin had 25 to 50 mm of clear liquid over the MLSS blanket and blanket surface is not even.

the influent rbCOD concentration is limited and not in excess of biological needs. Overmixing and aeration results in rbCOD being consumed by the microbes and further limiting rbCOD for driving denitrification and biological phosphorus removal.

3.2 Understanding Energy Requirement

Mixers operate continuously. Even those that are a small horsepower (hp) may contribute a significant amount of consumption to the electric bill. For example, a 7.5-kW (10-hp) motor operating continuously at 80% load, may consume more than 50 000 kWh/yr and cost approximately $4,000 in electricity. By using the correct energy input to match the process needs, a facility can make a significant step in reducing energy consumption.

Mixers with a rotating impeller create shear and flow. The equation for pumping by a mixer impeller is as follows (Oldshue, 1983):

$$Q = N_q \times rpm \times (Dia)^3 \tag{6.1}$$

Where Q = pumping rate at m^3/min;
rpm = revolutions per minute;
Dia = impeller diameter in meters, m; and
N_q = empirical impeller pumping characteristic, which varies with type of impeller and mixer manufacturer.

The importance of Equation 6.1 is that the pumping rate of a mixer is directly proportional to the mixer rpm. Reducing mixer speed from 100% to 50% reduces the pumping rate by 50%.

The second important equation relates to power consumption (Oldshue, 1983):

$$P = SPGR \times N_p \times (rpm)^3 \times (Dia)^5 \times C \tag{6.2}$$

Where P = power draw, kW;
$SPGR$ = specific gravity of water = 1 in most cases;
rpm = revolutions per minute;
Dia = diameter of impeller in millimeters, mm;
N_p = empirical impeller power draw characteristic, which varies with type of impeller; and
C = unit conversion factor = 2.15926×10^{20} (International Standard equivalent of C from Oldshue, 1983).

Equation 6.2 demonstrates that for a fixed-diameter impeller, reducing the rotational speed by 20% results in a 50% reduction in power use. If the speed is kept constant, and the impeller diameter is reduced by 20%, the power use will be reduced by 63%.

Figure 6.4 shows the mixer used in the study in which Figures 6.1 to 6.3 were taken. At full speed, the rated revolutions were 34 rpm at 3.7 kW (5 hp). Mixer pumping output and power draw is shown in Table 6.5. The pumping rates are a direct reduction linear with impeller rpm. However, the power draw is exponential with impeller rpm. A 17% speed reduction (83% of maximum rpm) decreases power draw by 44%. A 53% reduction in impeller rpm (47% of maximum rpm) reduces power draw by 90%. So, one option for energy efficient mixing operation is to add variable speed capability to the drive unit.

For the investment in a variable frequency drive (VFD), significant savings might be realized assuming that the initial mixing power input was more than needed by today's mixing criteria.

If installation of a VFD is not practical, then consider replacing the mixer impeller. Reducing the impeller diameter will reduce the mixer pumping rate (thrust produced) and mixer power draw. Table 6.6 shows the reduction in power draw and pumping rate for a range of impeller diameters turning at the same rpm.

FIGURE 6.4. Top entry mixer used in performance testing at the Rogers, Arkansas, WRRF

TABLE 6.5. Mixer pumping and power draw at lower impeller RPMs

Mixer, rpm	Pumping, lpm (gpm)	Pumping Rate Compared to Design, %	Power Draw, kW (hp)	Power Draw Compared to Design, %
34[a] (max speed)	143 700 (37 960)	100	3.7 (4.96)	100
28 (83% of max)	118 300 (31 260)	82	2.1 (2.77)	56
16 (47% of max)	67 600 (17 850)	47	0.39 (0.52)	10
8 (24% of max)	33 800 (8930)	24	0.05 (0.06)	1.3

[a]Manufacturer supplied data for 34 rpm operation. Pumping turndown and power consumption based upon calculations using Equations 6.1 and 6.2.

TABLE 6.6. Comparison of reducing impeller diameter on pumping and power draw

	Impeller Diameter	Impeller Pumping Rate	Impeller Power Draw	
Mixer, rpm	mm	lpm (gpm)	kW (hp)	% of Full Power
16	1829	76 820 (20 300)	0.477 (0.64)	100
16	1524	44 460 (11 740)	0.192 (0.26)	41
16	1219	22 760 (6010)	0.063 (0.084)	13
16	914	9600 (2540)	0.015 (0.02)	3.1

4.0 MIXING TECHNOLOGIES

There are many different types of mixers used in wastewater treatment. In general, mixers used in the liquid treatment process fall into two broad categories, top entry mixers and submersible mixers.

4.1 Top Entry Mixers

Figure 6.4 is an example of a top entry mixer. Top entry mixers have all parts except the shaft and impeller out of the liquid. This is a feature that operations/maintenance staff like, as the motor and gear box are easily accessible. The top entry mixers are mounted above the center of the tank on a platform or walkway. The platform or walkway used to support the mixer are typically not included in the cost of the mixer and the need to do so becomes a critical point in comparing the cost of different types of mixers. The top entry mixers are typically the more efficient mixing system. As described above, the mixing efficiency is primarily based on the pumping effect.

4.2 Submersible Mixers

Submersible mixers are positioned in the water so the mixer assembly is fully submerged. The submersible mixer is mounted on a mast or bracket. A winch is used to pull the mixer assembly out of the liquid for inspection and servicing. The mast for mixer mounting or wall bracket is significantly less expensive to install than a walkway or platform needed for a top entering mixer.

Submersible mixers come in many varieties from small diameter impellers turning at high rpm to the larger diameter impellers turning at a much lower speed. The larger diameter impellers turning at lower speeds are inherently more efficient than the small diameter high rpm units and can approach the efficiency of top entry mixers.

Small diameter impellers turning at a high-speed pump a higher velocity jet of liquid as compared to a top entry slow turning large diameter impeller. In liquid hydraulics, drag (friction) allows a relatively small volume stream flowing at an elevated velocity to eventually put a much larger volume of liquid in motion (mixing). The effect of drag can be seen when a jet of water from a hose enters a container of water. The liquid in the container begins to rotate in the direction that the hose is pointing. The higher velocity liquid discharged from the hose starts moving through the static liquid. Friction between the two fluids results in energy transfer from the higher energy moving stream to the static liquid. This friction, or drag, is an energy transfer of momentum energy.

The submersible mixer manufacturers report energy output as thrust. Thrust is the pushback from the mixer discharge that acts against the mounting mast. The thrust energy transferred into the mounting mast is measured and that thrust is the effective energy input into the tank of liquid from that submersible mixer.

Equations 6.1 and 6.2 are still valid for calculation of direct pumping and energy consumption from a high-speed direct drive mixer. The Nq and Np empirical constants are different for different impeller styles/configurations. The key for sizing submersible mixers is knowing the desired thrust for the mixing application. When switching from one type of submersible mixer to another, matching up the thrust output for each style of mixer is the way to get an equivalent comparison.

Mixer position is important for submersible mixers. Always contact the mixer vendor technical staff to define or verify mixer location and orientation. In some cases, computational fluid dynamic modeling may be necessary to define mixer location and orientation.

4.3 Selection of Top Entry Versus Submersible Technology

Based on the above information, a question evolves, as to why top entry mixers are not used for all mixing applications. The reason is installed cost. The detriment to top entry mixers is the cost to support and access (stairs, walkways, guardrails) the mixer. The platform or walkway may cost several times more than the cost of the mixer to be installed. So, for a top entry mixer that costs $X thousand, the cost of the platform or walkway may cost 5 to 10 times X, while for the submersible mixers, the cost of the mast is somewhere between ½ X to 1X times the mixer cost. Only in areas of the world where energy costs are $0.15/kWh or higher can the energy savings be enough to provide a reasonable payback period or return on investment.

Also to be considered are the maintenance costs of a submersible mixer, and the need to remove it from the fluid for maintenance. Doing so may require additional special equipment, or the need to ship the mixer off site for maintenance.

Therefore, the comparison between types of mixers must be a life-cycle analysis that accounts for the mixer, electrical equipment (including VFD if used), and installation hardware (mixer mast, platform, or walkway), and anticipated operating and maintenance costs.

Once the type of mixer is chosen for installation, there can still be an energy efficiency competition. There are many different types of mixers in the marketplace. All can provide the necessary mixing. However, some mixers are more efficient than others. For similar style impellers, the general principle is a larger diameter impeller turning at a lower speed is more efficient at pumping than a than a smaller diameter impeller turning at faster speed (higher rpm). Then there are the newer hydrodynamically designed impellers that pump more water more efficiently. There are typically several manufacturers providing similar products with which to have a competitive bid. Sometimes, the same mixer body may have more than one choice of impeller style to choose from.

For submersible mixers, the direct drive high-speed units will be the lowest cost mixers but will have a larger motor size than the gear drive submersible mixers. The gear drive units run at lower speeds, 100 to 300 rpm, and have larger diameter impellers. These are heavier units, because of the gear box, and often have a higher unit cost, but are more efficient at pumping. Finally, there are the banana blade mixers, named for the yellow composite material the impeller blades are made from. These are the 1.22- to 1.83-m (4- to 6-ft) diameter impellers, which rotate more slowly at approximately 50 rpm and pump a lot of liquid efficiently.

Figures 6.5 to 6.10 show some of the impeller styles and mixer body types that are available for use in wastewater applications. There are too many impeller blade configurations and mixer body types in the market to show them all and new products are entering the market all the time. When upgrading mixers, work with the local manufacturer or manufacturer's representatives to sort through reasonable options for your application.

4.4 Air Mixing

The preceding discussions have focused on mechanical mixing using a motor driven impeller. Big bubble mixers are now available in the marketplace. These big bubble mixers use large bubbles injected into a tank intermittently to provide mixing through airlift pumping action. The system consists of a number of diffusers or single pipe discharges spread around the basin in a geometric pattern. Each header is connected to small number of diffusers in a given area of the tank. Air is pulsed through the diffusers/single pipes to produce an airlift pump action. The air is pulsed for a short time and then shut off. The short time of aeration/blowing bubbles lifts liquid to the surface, creating a bottom to top liquid movement pattern just like a typical aeration system.

FIGURE 6.5. Direct drive submersible mixer in an anoxic zone application. 1.5-kW (2-hp) motor with a 305-mm (12-in.) impeller diameter and 550 rpm.

FIGURE 6.6. Direct drive submersible mixer providing supplemental mixing in an aeration basin

FIGURE 6.7. Flat blade turbine mixer in the original BNR basins at the Rogers WRRF (commissioned in the 1980s)

FIGURE 6.8. Closeup of the 1753-mm (69-in.) diameter impeller shown in Figure 6.4

A programmable logic controller controls operation of a number of solenoids. These solenoid valves cycle open and close to pulse air into various sections of the basin for a variable duration and defined mixing cycle. Not all solenoids are open at the same time but in a given cycle, air is pulsed to all areas of the tank to keep the liquid in motion.

The big bubble mixers are efficient at keeping solids in suspension but should not be used for rapid mixing (chemical feed) applications as they do not produce the shear needed for chemical dispersion. Sludge storage tanks, equalization tanks, and aerations zones that have a low air demand are some of the applications for these types of mixers. A single blower or air compressor can supply air to the mixing grid and these blowers will be much smaller than the blowers used in aeration of the tank volume. Typically, the blower/air supply is kept working continuously and air is distributed to the grids around the tank, which uses much less air than a typical aeration grid.

Pulsed Large Bubble Mixing has been used successfully in a number of mixing applications at WRRFs. The U.S. Environmental Protection Agency has identified this technology as an innovative and emerging energy conservation measure for anaerobic and anoxic zone mixing (EPA 832-R-10-005). It has been successfully applied in biological nutrient removal processes in

FIGURE 6.9. New hydrodynamic axial flow impeller design

northern climates. Concerns with oxygen transfer should be based on site conditions, wastewater characteristics, a flow and mass balance, and manufacturer's recommendations.

Some older and larger WRRFs may have a unique mixing challenge in mixing deep channels conveying return activated sludge (RAS) and a blend of RAS and primary effluent to the activated sludge basins. These channels can be 2.4 to 3.7 m (8 to 12 ft deep) and 1.2 to 3 m (4 to 10 ft wide) resulting in a geometry that is difficult to mix. Top entry mixers typically have an influence area of 2 to 3 times the basin width and submersible mixers have

FIGURE 6.10. Submersible large diameter impeller, slow speed mixer

an influence area of 3 to 5 times the width. For long channels, this results in a large number of small mechanical mixers to be installed.

Many times, a header either with coarse-bubble diffusers or simply holes drilled into a pipe, are used to distribute air along the length of the channel to keep solids in suspension. For channels more than 152 m (500 ft) in length, air use for mixing the channels can be a significant operating cost. Big bubble mixers could be a lower energy use alternative, but like the difference between submersible and top entry mixing, it may be necessary to perform a present worth analysis to balance equipment cost with energy consumption to select the most economical system.

For new facility designs, carefully consider channels or pipe systems using liquid velocity to keep solids in suspension. A better hydraulic design can minimize or eliminate energy use and should be a much easier system to maintain system.

4.5 Jet Aeration/Jet Mixing

Jet aeration is a unique type of a combination aeration and mixing system. It is typically found in sequencing batch reactor (SBR) applications and at industrial facilities. The MLSS is pumped/recirculated through a header under pressure and distributed back into the basin through nozzles. Air is distributed through a separate pipe and discharged through a shroud around the nozzle. The nozzles have a restricted opening to create back pressure, which accelerates the liquid, creating shear that can break up the air into coarse and fine bubbles depending on the air to liquid ratio.

In an SBR, the air will be turned off to create anoxic conditions and the liquid pumping will continue to provide mixing. The high-velocity liquid flow through the nozzle acts as an eductor to induce bulk liquid flow in the basin as described in Section 4.2. Jet mixing systems are sold separately, but the capital cost of the system and energy consumption may be higher than conventional top entry and submersible mixing systems.

4.6 Chemical Dispersion

Before BNR systems, rapid mixing was the typical WRRF mechanical mixing application. Rapid mixing is the dispersion of the chemical added into the main flow through the WRRF. The chemicals added can vary from ferric chloride or alum to chlorine solutions for disinfection to polymer addition. Each chemical has a different viscosity, and the more viscous the chemical, the more mixing energy needed for efficient distribution.

Typical design of a rapid mixing system is based upon a desired velocity gradient (G) value (power per volume ratio). Typically, to produce the high mixer energy draw, a radial pumping impeller is used. Radial pumping impellers are not efficient pumping impellers but they produce a lot of shear energy. The shear is needed to break up the viscous chemical being added and get it fully dispersed into the wastewater. Because rapid mix chambers are typically 30 to 60 seconds in retention time, the only major design factor is selection of the G value. Selection of the G value should be tailored to the viscosity of the chemical being added: lower G value for lower viscosity liquids and high G values for higher viscosity liquids. Inlet and outlet orientation in a rapid mix basin are also important. Intermittent mixer operation is not an option to reduce energy use. The chemical being added is typically critical to permit compliance and the cost of chemical use is typically higher than electricity consumption.

5.0 COMPARISON OF MIXING SYSTEMS

As stated throughout this section, there are factors other than direct cost that must be considered when selecting a mixer technology. Selection of the correct technology must be based on process requirements. The economic evaluation must be based on a life-cycle analysis. There are other intangibles that go into a selection, such as:

- Operational: Effect on other processes, equipment, and operator skill level.

  An example of mixing with air was provided. If substantial amounts of air are used for non-aeration process purposes, a parasitic load is placed on the aeration blowers that may make it difficult to control air distribution for biological treatment.
- Maintenance: The ability to maintain the proposed equipment.

  Submersible mixers may have a lower first cost than top entry mixers, but must be removed to be maintained. Depending on the size and weight, this may require more than one person and also require specialized equipment, perhaps a crane.

  Check on maintenance requirements from operators that have similar equipment.
- Warranty: The manufacturer's reputation for replacing equipment that breaks within the warranty period.
- Other equipment used on the site: The ability to use common technology and common parts reduces training and spare parts inventory. Standardizing on one mixer body but using different diameter or style of impellers is a possible approach.
- The ability to control mixing output.
- Noise
- Long-term experience

The life-cycle analysis must consider a number of parameters:

- Capital cost
  - Equipment cost
  - Equipment operating platform and associated structures
  - Walkways needed to reach the operating platform
  - Stairs needed to reach the walkways
  - Cost to supply the equipment with electrical power supply
  - Cost to supply the equipment with instrumentation and control
- Equipment operating requirements
  - Labor required to inspect

    - Labor required to operate
    - Electric energy requirements
- Equipment maintenance requirements
    - Cost of special tools
    - Cost of special equipment to remove the equipment, move, and load onto a truck for shipping to a maintenance facility
    - Cost of materials required for maintenance
    - The ability to maintain submerged bearings
    - Maintenance required to remove string caught on submerged rotating parts
- Other factors
    - Warranties
    - Chemical costs

6.0 REFERENCES

Kobylinski, E. A. (2014). Mixing myths revealed. *Influents Fall 2014 Volume 9 Official Publication of the Water Environment Association of Ontario*, pp. 29–32.

Oldshue, J. Y. (1983). Power correlations and the effect on mixing environment. In *Fluid Mixing Technology*. Chemical Engineering. McGraw-Hill Publications.

7.0 SUGGESTED READINGS

Bennett, C. O., & Myers, J. E. (1962). Chapter 1, The overall balances: Introduction to fluid behavior and Chapter 5, The overall momentum balance. In *Momentum, Heat and Mass Transfer*. Chemical Engineering Series. McGraw-Hill Publications.

ISO Standards: ISO 21630:2007.

EPA 832-R-10-005. Evaluation of Energy Conservation Measures for Wastewater Treatment Facilities, September 2010.

7

Aeration Systems

Tanja Rauch-Williams, Hong Yin, Ken Brischke, Eric Redmond, & Georgine Grissop

1.0 INTRODUCTION

This chapter covers energy considerations, efficiency, and optimization opportunities related to aeration systems. Specifically, this chapter includes a discussion of oxygen requirements, aeration equipment, oxygen transfer efficiencies, design and operational considerations, process monitoring and control, procurement that supports energy efficiency, and system optimization. Details on blowers are provided in Chapter 8 and additional details on instrumentation and controls are covered in Chapter 9. Chapter 12 includes a discussion of emerging aeration technologies.

2.0 DETERMINING OXYGEN REQUIREMENTS

2.1 Design Assumptions

In activated sludge treatment, oxygen serves as the electron acceptor to support the oxidation of organic and nitrogenous contaminants, bacterial synthesis, and decay of biomass. In aerobic activated sludge treatment, oxygen is used for organic carbon, ammonia, and hydrogen sulfide oxidation. Nitrification (conversion of ammonia to nitrite, and then to nitrate) requires oxygen, whereas denitrification (reducing nitrate to nitrogen gas) makes oxygen available. Denitrification also consumes carbon; thus, denitrification significantly lowers oxygen demand of nitrifying processes.

2.1.1 Liquid Stream Processes

Oxygen required to satisfy carbonaceous oxidation is site specific and typically ranges between 0.9 to 1.3 lb O_2 per pound of carbonaceous biochemical oxygen demand (cBOD) removed depending on the carbon composition in the secondary influent (Tchobanoglous et al., 2002). The oxygen demand for nitrification is 4.6 lb oxygen required per pound ammonia-N oxidized (Water Environment Federation [WEF], 2018). Denitrification makes half of this oxygen demand available (2.86 lb oxygen per pound nitrate-N removed). Aeration demand is also dependent on operational parameters, such as solids retention time (SRT) and mixed liquor suspended solids (MLSS).

The actual oxygen requirement (AOR) in liquid stream processes depends on several factors that are further discussed in this chapter, including wastewater flow and contaminant concentration, wastewater temperature, SRT, level and type of treatment, and basin and flow configuration.

2.1.2 Solids Stream Processes

At most larger facilities, sludge removed from liquid treatment processes is thickened, stabilized, and dewatered for beneficial reuse or disposal. Approximately 40% of all water resource recovery facilities (WRRFs) in the United States use aerobic digestion (WEF, 2017). Aerobic digestion uses aerobic bacteria to decompose organic matter and convert it into carbon dioxide.

Sidestream treatment processes that originate from solids processing are increasingly applied to either digested sludge (digestate) or recycle streams from dewatering post-anaerobic digestion. These processes are often aerated to oxidize the high ammonia concentrations before recycling back to the mainstream process. Aeration may be used in other processes such as for sludge aeration for intentional struvite precipitation.

2.2 Regulatory Design Criteria Requirements

General, rule-of-thumb regulatory requirements can be overly conservative, promoting designs and infrastructure that does not allow for energy-efficient aeration. As example, in the United States, many state agencies have established regulatory review and approval processes for the construction of new or modification of existing WRRFs. These states present specific design criteria guidelines or even design requirements for the design of wastewater treatment facilities that set forth specific design criteria for aeration systems and aeration demand estimates. These design requirements vary from state to state. For example, while some states follow the 10-State standards, other states such as Colorado, Texas, and Washington have updated criteria within the past 10 years and allowed design approaches for aeration systems that allow more flexibility for facilities to meet both low and high air demand requirements. Examples of such regulatory provisions include consideration of oxygen credits from denitrification, and guidance on supporting information needed for variance requests from recommended design criteria, for example for site-specific peak air demand.

Adequately sized and controlled aeration system have become more critical for WRRFs in the past decade, not only out of economic and environmental considerations, but because over-aeration can impede the removal of nutrients and total suspended solids (TSS) and compromise effluent quality. Many biological nutrient removal (BNR) facilities are not able with the current aeration systems to maintain desired target dissolved oxygen conditions throughout the process; the result is often increased effluent nitrogen and phosphorus concentrations.

2.3 Effect of Process Configuration

2.3.1 Liquid Stream Treatment

The oxygen demand in step-feed, complete mix, or plug flow processes is similar for achieving the same effluent quality, however the air demand may differ slightly. Step-feed and complete mix tank operation distributes the organic and ammonia load more evenly along the full length of the aeration basins. This can improve the distribution of airflow over a larger area of diffusers and result in improved oxygen transfer efficiencies. On the other hand, the alpha values typically increase along the length of plug flow reactors, improving the oxygen transfer efficiencies along the flow path.

2.3.2 Solids Treatment

The aerobic digestion process was once widely used to stabilize waste solids from municipal and industrial WRRFs. Compared to anaerobic digestion, aerobic digestion has high energy costs, reduced efficiency during cold weather, alkalinity depletion (which can be partially compensated through on/off aeration), and lower pathogen reduction. Over the past decade, the facility size at which anaerobic digestion can become economically defensible has decreased from approximately 20 mgd to less than 10 mgd. Although more expensive in construction, anaerobic digestion requires less energy input and biogas recovery can become a source of energy on- or offsite.

Autothermal thermophilic aerobic digestion gained popularity among small and midsized facilities. The process has a significant energy input for aeration of the digester and post-aerobic digester (SND tank) but achieves Class A biosolids.

2.3.3 Sidestream Treatment

Post-aerobic digestion, nitrification/denitrification, partial nitrification (nitritation), and anaerobic ammonium oxidation (anammox) are aerated sidestream treatment processes that can reduce the nitrogen recycled back to the liquid stream. Post-aerobic digestion involves the intermittent aeration of digestate post-anaerobic digestion to achieve nitrification and denitrification under low dissolved oxygen concentrations (Barber et al., 2019). Conventional nitrification/denitrification sidestream treatment operated at tapered and low dissolved oxygen concentrations can reduce energy input, recover alkalinity, and stabilize the pH.

Compared to conventional nitrification and denitrification, nitritation/anaerobic ammonium oxidation (anammox) is a sidestream process that requires less air for biological nitrogen removal. According to a full-scale sidestream deammonification study at Strass WRRF, nitritation/deammonification demonstrated an energy demand reduction of almost 80% for nitrogen removal compared to conventional nitrification and denitrification process (1.2 kWh/kg N [0.7 hp-h/lb N] for nitritation/deammonification) as opposed to 5 to 6 kWh/kg N [3.0 to 3.6 hp-h/lb N]) (Wett et al., 2010). Limitations of sidestream deammonification identified by some facilities may include lack of process stability, retainment of anammox granules, equipment vulnerabilities (design dependent), and limitations of proprietary technology controls. When selecting energy-efficient sidestream removal processes, the overall energy demand of the system (not just the of the process) need to be considered, which may include mixing and pumping. As of today, this process has been successfully implemented by more than a dozen facilities in the United States in various configurations.

Conventionally, ammonia is converted to nitrite and nitrate via autotrophic nitrification and subsequently reduced to nitrogen gas through heterotrophic denitrification. This route demands energy (4.57 lb O_2/lb NH_4-N oxidized) and 4.8 to 6 lb of organic carbon per pound of nitrogen that is not available for energy recovery (e.g., in form of biogas). Nitrite shunt and partial nitritation/deammonification (anammox) processes require stoichiometrically less oxygen and less organic carbon for the same amount of nitrogen removal. Nitritation/denitritation (Nit/DNit) saves 25% of the total oxygen demand, 40% of the carbon demand, and 50% of sludge production compared to nitrification and denitrification (Turk and Mavinic, 1987).

The partial nitritation/deammonification processes removes ammonia by partially nitrifying ammonia to nitrite followed by anammox bacteria converting nitrite and remaining ammonia to nitrogen gas. This process requires 63% less oxygen, and nearly 100% less organic carbon (van der Star et al., 2007), 80% less biomass is produced; however, the process requires more carbonate alkalinity compared to nitrification/denitrification despite the fact that only half of the ammonia is oxidized.

2.4 Processes with Lower Oxygen Demand

2.4.1 Simultaneous Nitrification/Denitrification

Processes that remove nitrogen without dedicated anoxic zones (lack of dissolved oxygen) are commonly believed to accomplish this through so-called simultaneous nitrification/denitrification (SNDN). In fact, this may be a simplification and several different processes may happen in parallel under low dissolved oxygen conditions resulting in the removal of ammonia and nitrogen.

Simultaneous nitrification/denitrification can reduce the overall oxygen demand through the recovery of oxygen during denitrification. A number of facilities operate under SNDN by maintaining low dissolved oxygen concentrations. Simultaneous nitrification/denitrification is a prevalent process in oxidation ditch processes.

2.4.2 Low Dissolved Oxygen Nitrification

It is still commonly believed and reflected in most state design criteria that complete nitrification requires dissolved oxygen concentrations of at least 1.5 to 2 mg/L or higher. However, facilities exist that achieve full nitrification and almost complete denitrification year-round even in cold temperatures with dissolved oxygen concentrations never exceeding 0.3 to 0.5 mg/L. These operations are consistent with Downing and Scragg (1958), who reported that a dissolved oxygen of at least 0.3 mg/L was needed for nitrification to occur. Microbial adaptation and overall stability of process conditions play a critical role in achieving low dissolved oxygen nitrification. Use of ammonia-based aeration control, using ammonia probes to control airflow, have also shown this to be factual.

2.4.3 Solids Treatment

Post-aerobic digestion is a digestion process in which aerobic digestion is placed after an anaerobic digestion step, thus partially denitrifying ammonia and improving volatile solids (VS) removal (Johnson et al., 2011). Operation of the post-aerobic digester using either cyclic aeration or simultaneous Nit/DNit at low dissolved oxygen provides total nitrogen removal without the need for external carbon source addition. Alkalinity addition may be required to stabilize the process operation. Greater than 90% ammonia removal has been observed without the need for chemical addition and near complete total nitrogen removal can be achieved if Nit/DNit at low dissolved oxygen concentrations can be optimized (Johnson et al., 2011).

Intermittent aeration during aerobic digestion and post-aerobic digestion reduces oxygen demand and provides the additional benefit that alkalinity is recovered through denitrification of nitrate. This can help with pH stabilization in the digesters. Including anoxic zones in the design or on/off cycling of air can recover up to 50% of the alkalinity and reduce oxygen requirements by as much as 18% (Daigger and Bailey, 2000).

3.0 TYPES OF AERATION EQUIPMENT

Aeration exposes wastewater to oxygen so that it can be transferred into the fluid. A primary use for aeration (oxygen transfer) is to provide oxygen to microorganisms in wastewater treatment. Aeration is further used to provide for mixing of tank contents, in aerated grit chambers, or to increase dissolved oxygen in the treated water before discharge (post-aeration).

Diffused aeration exposes air to water or wastewater. Mechanical aeration exposes water or wastewater to air. Increasing the gas–liquid interfacial area increases the oxygen transfer rate.

A variety of devices have been developed and used to accomplish aeration. These have been discussed at length in references such as *Design of Municipal Wastewater Treatment Plants* (WEF, 2018) as well as *Wastewater Engineering, Treatment and Reuse* (Metcalf & Eddy, 2014). This section provides a brief overview of aeration devices with focus on their respective energy efficiency. For comparative purposes, oxygen transfer efficiency is typically identified as the Standard Aeration Efficiency (SAE), which is the mass of oxygen transferred divided by the energy required to transfer said oxygen mass over a specific time period; kgO_2/kWh (lbO_2/hp-h). The SAE's for surface and submerged aerators are summarized in Table 7.1.

3.1 Surface Aerators

3.1.1 Low-Speed Aerators

A low-speed aerator consists of a large-diameter impeller mounted on a vertical shaft that splashes the mixed liquor from an aeration basin into the air. These aerators are typically used in older complete mixed activated sludge facilities or oxidation ditch systems. Oxygen is transferred from the air into the water droplets as they come in contact. Power draw increases significantly even with small increases in the impeller submergence depth. To optimize oxygen transfer and power consumption, oxidation ditches have been built with motor-operated weirs that vary impeller submergence based on the desired dissolved oxygen concentration in the basin. However, in some cases, these systems have required too much time and expense to maintain and thus eventually were removed or abandoned in place.

3.1.2 High-Speed Aerators

High-speed aerators are often associated with lagoons (floating aerators) and oxidation ditches (brush aerators). Lagoon aerators are typically mounted on floats held in place by guy wires anchored at the edge of the lagoon. The impeller splashes the wastewater into the atmosphere, resulting in oxygen transfer from air into the water droplets. Aeration is controlled through an operator-adjusted timer depending on daily or seasonal flow and load variation. The float rises and falls with the liquid depth in the lagoon; the submergence of the impeller is relatively constant.

Brush aerators are most commonly associated with oxidation ditches. Oxygen is transferred from the air to the water droplets. At the same time, the tines impart a velocity to the water, moving it around the oxidation ditch. As with low-speed aerators, the energy draw on the motor increases with small increases in liquid depth resulting from flow increases. This also results in an increase in oxygen transfer. Oxidation ditches have been constructed with motor-operated, adjustable weirs to allow the operator to reduce aerator submergence.

3.1.3 Disk Aerators

Disk aerators are a variant of the brush aerators used in oxidation ditches. Instead of a drum with tines, the aerator consists of an axle with disks. As the rotor turns, the disks impart velocity to the water. Some of the water is splashed into the atmosphere and absorbs oxygen.

3.2 Submerged Aerators

3.2.1 Diffusers

Diffused aeration delivers compressed air through porous or perforated diffusers. Oxygen is transferred from the air bubbles to the bulk fluid. Diffusers are categorized as coarse bubble (typical bubble size >5 mm [>0.2 in.]) or fine-pore (bubble size <5 mm [<0.2 in.]). Both types are described in the following subsections.

3.2.1.1 Coarse Bubble

Coarse-bubble diffusers were commonly used for aeration from the 1950s through the 1970s. The energy crisis of the 1970s resulted in reevaluation, development, and application of fine-pore diffusers. Today, coarse-bubble diffusers are primarily used at small WRRFs, in anoxic selector zones, for mixing in channels, and in aerobic digesters. These diffusers are typically fabricated of heavy gauge polyvinyl chloride (PVC) or thin gauge stainless steel. Large holes are punched along the sides to give a wide airflow range.

3.2.1.2 Fine Pore

Fine-pore diffusers generate smaller air bubbles, increasing the interfacial area between air and water, resulting in higher oxygen transfer rates. The higher transfer rates reduce the amount of air needing to be compressed, thereby reducing the cost of operating aeration blowers.

Fine-pore diffusers are most commonly made of porous ceramic material or perforated elastomeric membrane material. Airflow to ceramic diffusers must be maintained, even when a basin is off line, or mixed liquor may enter the pores and foul the diffusers. Membrane diffusers are available as circular inserts to a disc holder, horizontal tubes, or panels installed on the aeration basin floor. Without aeration, perforations in the membrane diffusers will close such that the diffuser is protected from fouling. Disc diffusers are commonly mounted on PVC pipes and arranged in a grid pattern across the aeration basin floor. Tube diffusers are generally available in a variety of lengths and can be installed in a greater density than discs. Tube diffusers can typically be mounted closer to the basin floor, which can be useful in achieving desired oxygen transfer efficiency (OTE) when retrofitting diffusers in an existing shallow aeration basin. Panel diffusers are efficient because they can cover more than 50% of the floor area, compared to an upper limit of approximately 25% for circular disc diffusers. Panel diffusers are available in both wide and narrow (strip) configurations.

Fine-pore diffusers generally have a relatively narrow operating range in which uniform airflow and bubble pattern across the surface can be achieved. It is important to note that fine-pore diffusers made from flexible membranes can produce bubbles of varying diameter depending on the airflow. Thus a "fine-pore" membrane diffuser can produce "coarse bubbles" (>5 mm [>0.2 in.] diameter) at high flow. In comparison, rigid-pore fine-bubble diffusers (ceramic) tend to produce "fine bubbles for a wider range of air flows." See U.S. Environmental Protection Agency (U.S. EPA) *Design Manual for Fine Pore Diffusers* for more detailed information on different diffuser materials (U.S. EPA, 1989). The reader is advised to work with the supplier of the diffusers that will be used to obtain diffuser specific information.

The suggested life expectancy of membrane diffusers is approximately 10 years, although diffusers at some facilities exceed this threshold. Ceramic diffusers have an expected life of more than 20 years. Membrane diffusers can be manufactured from a variety of different materials. When replacing existing diffusers with diffusers of a different material, care should be taken to ensure the headloss across the new diffusers is the same as those being replaced to ensure the operating pressure of the existing blowers is not exceeded. As with ceramic diffusers, membrane diffusers are subject to fouling that deteriorates OTE; periodic flexing is reported to mitigate biological fouling. The expected life for membrane diffusers depends on the conditions of operation and frequency of regular cleaning.

3.2.1.3 Jet Aerators Pore

Jet aerators use water jets operated by pumps outside the aeration tank to circulate wastewater and mix the tank contents. Compressed or ambient air is introduced from a separate pipe attached to the water pipe and ejected through a nozzle assembly. As the air comes out, it diffuses into small bubbles, causing oxygen transfer from air to water.

3.2.1.4 Sparger Turbines

Sparger turbines are a combination of mechanical aerator and diffused aeration. The most visible part of this aerator is the motor and associated gear box to drive a vertical mixer, mounted at some depth below the water surface. Just below the mixer impeller is a sparger ring. Air is introduced through a blower and discharged through the sparge ring below each mixer. As air bubbles pass through the impeller, they are sheared into smaller bubbles and dispersed throughout the tank. Sufficient air is provided to meet oxygen requirements and the turbine is used for mixing.

3.2.1.5 Diffuser Submergence

The diffuser submergence is the major contribution to the operating blower pressure, while pressure losses in the aeration pipes and appurtenances and the diffuser head are minor in comparison. Maximizing the diffuser submergence in coordination with the blower equipment and efficiency curves increases the OTE (of approximately 1% to 2% per foot) and thus can save energy. Special air pipe designs on the floor of aeration tanks can allow placing the laterals at a lower elevation than the centerline of the air header, thereby increasing diffuser submergence further.

3.3 Standard Aeration Efficiency of Aerator Systems

The SAE is a measure of the mass of oxygen transferred into clean water per unit of power over a specific time period, typically, kilograms of oxygen/kilowatts per hour. A summary of commonly referenced SAE for the aeration devices is listed in Table 7.1.

4.0 MIXING

Maintaining minimum mixing energy in aerated reactors is an important consideration of aeration system designs to keep solids in suspension and distribute oxygen uniformly throughout the reactor. Maintaining sufficient velocities throughout the reactor and on the bottom of the basins minimizes solids deposition. What the minimum required mixing energy is that should be provided in a given aerated basins zone remains a topic of debate. Lately, traditional rules of thumb for industry recommended mixing energy

TABLE 7.1. Typical SAE and airflow weighted α factors for commercially available aeration systems (IWA, 2018).

Aerator Type	SAE kgO_2/kWh ($lbO_2/hp\text{-}h$)	SAE @ Low SRT at 2 mg/L Dissolved Oxygen kgO_2/kWh ($lbO_2/hp\text{-}h$)	SAE @ High SRT at 2 mg/L Dissolved Oxygen kgO_2/kWh ($lbO_2/hp\text{-}h$)	Alpha Factor
Surface Aerator—High Speed	0.9–1.3 (1.5–2.1)	0.4–0.8 (0.7–1.3)		0.47–0.64
Surface Aerator—Low Speed	1.5–2.1 (2.5–3.5)	0.7–1.5 (1.2–2.5)		0.48–0.71
Turbine	1.2–1.8 (2.0–3.0)	0.4–0.6 (0.7–1.0)	0.6–0.8 (1.0–1.3)	0.30–0.47
Jet	1.5–1.7 (2.5–2.8)	0.8–1.2 (1.3–2.0)		0.52–0.71
Coarse Bubble	0.6–1.5 (1.0–2.5)	0.3–0.7 (0.5–1.2)	0.4–0.9 (0.7–1.5)	0.50–0.64
Fine Bubble	3.6–4.8 (5.9–7.9)	0.7–1.0 (1.2–1.6)	2.0–2.6 (3.3–4.3)	0.3–0.65

guidelines have been challenged as being overly conservative. See Chapter 6 for a discussion on mixing design criteria using aeration.

In particular, mixing requirements need to be considered in the design of the diffuser system in the downstream zones of plug flow reactors where the aeration demand is low. Because of the higher efficiency of fine-bubble diffusers, the airflow for the later aerated zone(s) of an aeration tank can be dictated by the required mixing energy and recommended minimum airflows per diffuser, and not oxygen transfer requirements.

Oxidation ditch-type aeration systems that are being converted to fine-bubble diffused aeration, however, can become mixing limited. Additional energy is required to maintain the required horizontal velocity in the ditch to keep solids in suspension. Horizontal mixing has also been demonstrated in field testing to increases the OTE of the diffuser system. Thus, a detailed analysis is required to coordinate mixing and aeration systems to determine the optimum efficient overall system.

Installing variable frequency drives (VFDs) on surface brush aerators in oxidation ditches and controlling the speed of rotation based on dissolved oxygen readings is one way to adjust aeration and save energy. Lower rotation speeds may limit mixing for keeping solids in suspension. The published energy input required in oxidation ditches for mechanical mixing ranges anywhere from 1 to 40 kJ/s·ML (5 to 203 HP/mil. gal) reactor volume. Rather than relying on empirical values, it is recommended to use computational fluid dynamics (CFD) modeling to assess the optimal mixing energy, location, and orientation that depends on reactor geometries and other factors.

5.0 OXYGEN TRANSFER EFFICIENCY

5.1 Diffuser Flux Rate

Diffusers are generally used in subsurface system to introduce air into the wastewater. The two most common aeration systems are subsurface and mechanical aerators. Fine-pore diffusers are the most popular diffusers used in the United States, which have gained increased interest over mechanical aerators since the 1970s, because of higher oxygen transfer efficiencies.

For fine- or coarse-bubble diffusers, the diffuser air flux (airflow per diffuser unit area) is dependent on the diffuser type, design configuration, and airflow. Higher diffuser flux in membrane diffusers results in larger bubbles because of coalescing, and for membrane diffusers, larger pore size is caused by membrane stretching. Larger bubbles result in reduced OTE. Lower diffuser flux results in smaller bubbles and increased OTE. These relationship curves are specific to each diffuser and can be requested from manufacturers.

5.2 Alpha Factor

The α ("alpha") factor combines a number of system characteristics that affect the site-specific oxygen transfer rate related to wastewater source (industrial discharge to municipal WRRFs) and wastewater characteristics, aeration tank layout, and aeration system equipment. Aerators are typically tested in clean water to define the oxygen transfer rate and OTE of specific equipment under standard conditions (SOTE). The SOTE of aerators is greater in clean water than in wastewater and the α factor is the correction factor to account for lower transfer efficiency in wastewater under actually installed field conditions.

Values for α range from approximately 0.3 to 1.0 (Tchobanoglous et al., 2002), depending on the type of aeration equipment used. The α factor is directly proportionate to the OTE under field conditions and thus to the energy input of aeration equipment. Ways to increase the alpha under field conditions include

- plug flow basin configuration,
- lateral mixing (e.g., as provided in oxidation ditches),
- unaerated selectors basins,
- operation at elevated SRT,
- operating at lower MLSS, and
- clean diffusers periodically to minimize fouling.

5.3 Air Demand Calculations

5.3.1 Liquid Stream

Oxygen required for the biological process, known as the actual oxygen requirement (AOR), should meet the demand for cBOD removal and total nitrogen removal. Standard oxygen requirement (SOR), which is the amount of oxygen to be transferred under standard condition to meet the AOR under field conditions, can be estimated from the AOR after adjusting for the environmental conditions in the aeration tank. The air demand can then be calculated using the SOR divided by the SOTE of the aeration equipment.

Air demand calculation should also consider air requirement to maintain biomass in suspension (see previous section).

Process aeration demand depends on operational parameters, such as SRT, MLSS, dissolved oxygen concentration, wastewater characteristics and temperature, diffuser conditions, and so forth. Processes operated at long SRT require more oxygen because of endogenous respiration of mixed liquor volatile suspended solids (MLVSS). Short SRTs may decrease the alpha factor and thereby increase oxygen demand because of lower oxygen transfer efficiencies in presence of organic pollutants, such as detergents. Oxygen transfer rate decreases as mixed liquor dissolved oxygen concentration increases. The dissolved oxygen concentration to maintain aerobic floc varies with the specific facility. Although traditionally dissolved oxygen concentrations above 1.0 ppm have been considered adequate, several first pilot studies and full-scale facilities have now demonstrated that full nitrification can be achieved at dissolved oxygen concentrations as low as 0.2 mg/L (Fitzgerald et al., 2015; Hanaki et al., 1990; Keene et al., 2017). Anaerobic or anoxic selectors are able to further increase aeration efficiency as they allow that readily available substrate sorbs on flocs to be metabolized, which otherwise would accumulate at bubble surfaces and decrease aeration efficiency.

Nitritation/denitritation, and nitritation/deammonification present significant savings in aeration power, supplemental carbon, and alkalinity use. Table 7.2 below summarizes these benefits compared to traditional nitrification and denitrification process (Bowden et al., 2016).

5.3.2 Solids Stream

High solid concentrations above 1% total solids required high energy for mixing and aeration system in aerobic digesters because of high viscosities and high oxygen uptake rates in particular during the initial 1 to 2 weeks of digestion. The aerators and diffusers used for aerobic digesters are the same as those used for secondary aeration. Aerobic digestion is limited by mixing and oxygen requirements. Generally speaking, mixing requirements for digesters with mechanical aeration can range from 2 to 4 kW/100 m^3 (0.76 to 1.5 hp/1000 cu ft).

TABLE 7.2. Benefits of energy-efficient aeration process

	Nitrification/ Denitrification	Nitritation/ Denitritation	Nitritation/ Deammonification
Air demand (g/g NH4-N oxidized)	4.57	3.42	1.9
Carbon demand (g COD/g N removed)	3.5–4.0	2.0–2.5	0.5
Energy (kWh/kg N removed)	4.0–6.5	2.3–2.9	1.1–1.7

The calculation of AOR and SOR are explained in detail in U.S. EPA's *Design Manual: Fine Pore Aeration Systems* (U.S. EPA, 1989).

Coarse-bubble diffused air systems require approximately 0.9 to 1.8 m^3/s per 100 m^3 (19 to 28 cfm/1000 cu ft). Oxygen requirements can vary widely from facility to facility. The reduction of cell tissue requires approximately 2.3 kg O_2/kg of VS (2.3 lb O_2/lb of VS) destroyed. If primary sludge is included in the aerobic digestion process, it requires an additional 1.6 to 1.9 kg O_2/kg BOD (1.6 to 1.9 lb O_2/lb BOD) removed by the primary clarifiers. The air requirements for aerobic digestion can equal or exceed the air required for secondary aeration.

In the past, fine-pore diffusers were considered unacceptable for digesters because of concerns over fouling of the diffusers. Recent practice has shown that fine-pore diffusers can provide reliable operation in digesters. The superior mixing and oxygen transfer potential of fine-pore diffusers make them the preferred technology for most systems. They should be considered as an energy conservation measure for existing aerobic digesters that have either mechanical aeration or coarse-bubble diffusers (Jenkins, 2013).

6.0 DESIGN CONSIDERATIONS

Designers have an obligation to include provisions that provide WRRFs with the ability and flexibility to operate energy-efficient aeration systems throughout the life span of their facilities. The following briefly discusses design considerations that should be considered for aeration tank and aerator system designs.

6.1 Supplemental Mixing

Using aeration systems for the dual purpose of tank mixing and oxygen transfer may not only result in inefficient energy designs but may also limit process performance. Over-aeration, in particular at the end of aeration tanks before final clarification, can result in shearing of flocs, pin floc formation, and resulting elevated effluent TSS concentrations. Operators may not have the ability to optimize aeration for optimal removal of nutrients (nitrogen and phosphorus) separate from allowing for sufficient mixing. Decoupling aeration from mixing systems in flow through aeration processes, oxidation ditches, or other process configurations requires more capital cost but can pay off through energy savings. There are a variety of cost-effective approaches, including supplementing mechanical mixing, nested layouts of multiple fine-bubble diffuser grids, or nested layouts of fine-bubble and large-bubble diffuser grids. Facilities with the potential of benefitting from supplementing mixing are encouraged to evaluate options for retrofitting their facilities.

6.2 Aeration System Design

6.2.1 Phased Aeration System Designs

For most facilities, a key challenge is to design aeration system appropriately to accommodate flow and load conditions anticipated under process startup and ultimate design conditions. Often, this span is too long to provide adequate and energy-efficient designs. As a result, design engineers default to recommendations that satisfy design conditions and cost considerations. As a result, facilities may be left with equipment that is oversized and designed for air and energy wastefulness from the outset of operation.

6.2.2 Diffuser Layout Considerations

Tapering diffuser densities along the aeration basin zones in coordination with anticipated oxygen demand (loading) allows operating diffusers closer to the optimal air fluxes for energy efficiency. These are to be coordinated with the respective diffuser manufacturer.

Addition of blank diffuser plugs or caps in designs and specifications is a low-cost option to allow operation with a reduced number of diffusers initially while allowing flexibility to increase the number of active diffusers in future years as needed. At minimum, it is good practice to include 10% of diffuser blanks, but the fraction could be much higher in case of large differences between startup and design airflow demands.

In special situations when design airflows need to span a wide range, interlaced diffuser grids can allow operators to reduce the active diffuser density by half or two thirds simply by shutting off separate valves for each interlaced grid.

Operating in step-feed or complete-mix modes may result in a more uniform overall alpha by distributing surfactants throughout the aeration tank and allowing for their dilution and more rapid removal by adsorption (U.S. EPA, 1989) (see earlier discussion in this chapter). Another possible benefit for a fine-pore system is the reduction of diffuser fouling, which is known to occur at the inlet-area diffusers in plug-flow systems.

6.2.3 Swing Zones

Designs should consider aerobic/anoxic swing zones, which are dedicated zones equipped with both diffusers and mixers. These zones provide operators with flexibility to adjust the aerobic volume in reactors seasonally or with load changes over time to best meet desired effluent quality objectives.

6.2.4 Air Valves

The design of aeration valves needs to be closely coordinated with the airflow requirements and all other aeration system components over all operational conditions. The type of valve, the number of diffuser grids controlled manually or automatically, valve sizing and location along the pipeline, as well as control logic are all critical factors that determine the accuracy of aeration control and energy efficiency (see Chapter 9 for more detail).

7.0 OPERATIONAL CONSIDERATIONS

7.1 Target Dissolved Oxygen Concentrations

The most common cause of aeration system inefficiency is greater than necessary dissolved oxygen concentration in the mixed liquor, which is generated at the expense of high-energy use. The closer the dissolved oxygen is to saturation, the greater the resistance for dissolution of oxygen will be and, hence, the lower the oxygen transfer rate. Oxygen transfer from air to water is a mass transfer reaction, directly proportional to the driving force, which is the difference between the dissolved oxygen saturation concentration and the dissolved oxygen in bulk liquid. To optimize OTE, the operator must operate the aeration tank at the minimum dissolved oxygen concentration to achieve process objectives. Acclimating the biology to low dissolved oxygen operation while still reliably achieving full nitrification, has a higher success rate when operators first focus on stabilizing dissolved oxygen concentrations throughout the aeration basins, by reducing dissolved oxygen spikes that typically occur during low flow and load conditions (e.g., at nighttime) and in the downstream sections of plug flow reactors. Once the biology is exposed to a diurnally stable dissolved oxygen environment, operators can start to reduce dissolved oxygen concentrations patiently in a slow and controlled matter to properly acclimate the biology.

When manual control is practiced, operators will tend to overcompensate to "keep ahead of the game" should an increase in load occur before the next mixed liquor dissolved oxygen reading. Installation of online dissolved oxygen and nutrient probes coupled with online aeration control systems can help avoiding the occurrence of excessive mixed liquor dissolved oxygen and help reduce overall dissolved oxygen concentrations while safely meeting effluent requirements. Evolution in the technology for online metering and controls has resulted in minimum maintenance to achieve reliable results (see Chapter 9 for further details). In medium-to-large WRRFs, the payback period for installing automated blower control from mixed liquor dissolved oxygen probes is generally within a few years.

Dynamic dissolved oxygen setpoints that vary with diurnal loading (higher dissolved oxygen during high loading, lower dissolved oxygen during low loading) can produce aeration savings between 10% and 30% compared to fixed dissolved oxygen setpoints based on the experience of the authors at full-scale facilities. The dynamic dissolved oxygen setpoints can be controlled by using online ammonia analyzers in a feedback or feedforward control loop (feed forward is preferred over feedback as feedback may miss the peak loadings) (see Section 7.2 and Chapter 9 for more detail). Dynamic aeration control implementing the combination of ammonium and nitrate monitoring to increase/decrease dissolved oxygen concentrations provides increased airflow savings and can divert the use of carbon for nutrient removal or energy recovery.

7.2 Selection of Online Probe Locations

The location of the dissolved oxygen probes affects the ability of the system to identify and respond to changes in demand for oxygen in a timely manner. If located at the end of a plug-flow aeration basin, the dissolved oxygen sag caused by a peak loading would not be observed until well after the peak has entered the tank.

In facilities with multiple-pass aeration basins, two dissolved oxygen meters may be needed for each pass located at the inlet and outlet to adequately tailor air supply. Hand-held dissolved oxygen probes can be used to ensure that all the basins are operating in a similar manner. In facilities requiring nitrification, placement of online ammonia probes at strategic locations along the flow path in aeration basins provides early indication of ammonia breakthrough and potential for aeration reduction.

Technologies have been developed that allow for online offgas analysis to determine the actual update rate of oxygen supplied to a given basin area. These systems require careful setup, calibration, monitoring, and adjustment to field conditions but can help to better match air supply to process oxygen demand.

The return of investment of using online probes and analyzers for distributing oxygen to locations should not only be measured in terms of reduced power costs and greenhouse gas reduction; the ability of operators to tightly control aeration is critical for achieving the maximum potential of biological nutrient removal.

7.2.1 Process Monitoring and Control

Most WRRFs have characteristic daily, weekly, and seasonal loadings. With the appropriate analysis and study and some around-the-clock sampling and analysis of aeration basin influent and sidestream recycle flows, the wastewater can be characterized and air supply optimized to match diurnal variation.

7.3 Adequate Number of Tanks in Service

Often, facilities will operate all parallel tanks in service even when flows and loads would allow taking tanks out of service for extended periods of time. Assuming the SRT is held constant, operating more basin volume than needed results in excess energy demand for aeration and mixing.

Concerns about the upkeep of mechanical equipment and degrading diffuser conditions can be managed through tailored maintenance plans (e.g., rotating or alternating operation).

7.4 Dissolved Oxygen Management in Membrane Bioreactor Systems

In addition to biological process aeration, membrane bioreactor (MBR) systems require air for scouring of the membranes. In recent years, several MBR manufacturers have developed new technologies that allow lowering the energy demand for air scour in MBR systems. The energy consumption of MBR systems is therefore greater than of a conventional activated sludge processes and more than 50% of the energy used for MBRs' treatment is for aeration. Barillon et al. (2012) reports that membrane scouring, biological aeration, and sludge circulation are three major elements for energy consumption and contributed 23% to 58%, 18% to 40%, and 9% to 22% of total energy consumption of MBR systems.

Module design, membrane performance, effluent quality, membrane utilization, and operational strategies are factors determining the energy use. Reducing fouling through membrane scouring and clean-in-place systems, and energy-efficient diffusers for cleaning can be design elements of energy-efficient MBR systems. The membrane technology supplier may define the frequency and airflow required for membrane scouring. The design of the membrane aeration system is also specific to the membrane supplier, which is typically supplied along with the membrane units.

In practice, the dissolved oxygen in the membrane tank and return activated sludge (RAS; recycled typically at ratios of 300% to 500% of influent flowrate) can be as high as 5.0 mg/L. By designing a suitable RAS feeding point within the process train, it is feasible to use the residual dissolved oxygen to optimize the total aeration requirements of the process. For example, when a Modified Ludzack-Ettinger (MLE) process is applied to treat municipal wastewater with the goal of achieving effluent total nitrogen of less than 10 mg/L, the RAS can be fed to the aerobic reactor to use the residual dissolved oxygen in the mixed liquor. Nevertheless, an additional internal biomass recirculation for nitrates, between the aerobic reactor and the anoxic reactor, would be required for denitrification. If low effluent total nitrogen (e.g., 5 mg/L) with or without external carbon is required through the MLE process, then a de-oxygenation and denitrification zone can be added within the RAS loop for use of the residual dissolved oxygen by endogenous respiration and partial denitrification. Subsequently, RAS without dissolved oxygen is fed to the anoxic reactor to maximize denitrification and accomplish low-effluent total nitrogen concentration.

Effluent permits often specify dissolved oxygen requirements (e.g., 5 to 6 mg/L) and, because MBR effluent (permeate) has similar levels of residual dissolved oxygen, the energy for post-aeration is minimized.

7.5 Diffuser Cleaning

Ongoing maintenance is required for fine-pore diffusers to maintain OTE as well as operating pressure within reasonable limits. This applies to both ceramic and membrane diffusers. The manufacturer of the installed equipment should be contacted for specific guidelines.

Bio-fouling can also occur when diffusers are operated below their recommended minimum flux rate. Insufficient airflow does not produce enough activity on the membrane to inhibit biological growth on the membrane material itself.

A higher operating pressure from diffusers results in an increased power consumption by the blowers. As a simplified example, an increasing operating pressure of 0.5 psi because of diffuser fouling results in an increased blower power consumption of 40 J/s per ML (0.2 HP per mil. gal. per day) treated. At a blower efficiency of 75%, motor efficiency of 92%, and VFD efficiency of 95%, this would result in an increase of 2020 kWh/yr (2710 HP-h per year) for each million gallon per day of flow.

Evaluating diffusers for fouling is particularly critical after an operational upset. For example, a loss of air to the diffusers (especially ceramic diffusers) could result in intrusion of mixed liquor solids, causing the diffuser pores to become fouled.

In situ methods for water-side cleaning are available for either full or drained tanks. In situations involving dewatering aeration tanks, personnel safety is paramount. Safety concerns when performing in situ cleaning of diffusers should be addressed. Provisions must also be made to reduce the risk of damage to the diffuser system from exposure to direct sunlight or excessive thermal expansion and contraction.

In situ methods include physical, chemical, and biological procedures. The diffusers may be physically washed with water, air, or steam, which dislodges loose external biological growths and deposits (check with the manufacturer for their recommendations). Chemical procedures may include addition of manufacturer approved gaseous compounds such as formic acid, hydrogen chloride, chlorine, or gaseous biocides to the air side or may include liquid acid or detergent cleaning of diffuser surfaces. Figure 7.1 shows a mobile acid diffuser cart. Ex situ methods are often cost-prohibitive today, including refiring of ceramic diffusers; high-pressure water jetting; and washing ceramic diffusers with silicate-phosphorus, alkali, acid, or detergent. Figure 7.1 shows a mobile diffuser cart.

Some manufacturers of perforated membrane diffusers suggest weekly or monthly air bumping, also called flexing. Porous stone (ceramic) diffusers have been cleaned in situ by means of acid gas injection. This involves the injection of hydrogen chloride or formic acid gas into the air delivered to diffusers. Safety training and precautions are essential. Gas injection typically continues until the pressure differential across the diffusers has stabilized, and the pressure stabilizes after approximately 30 minutes of acid gas flow.

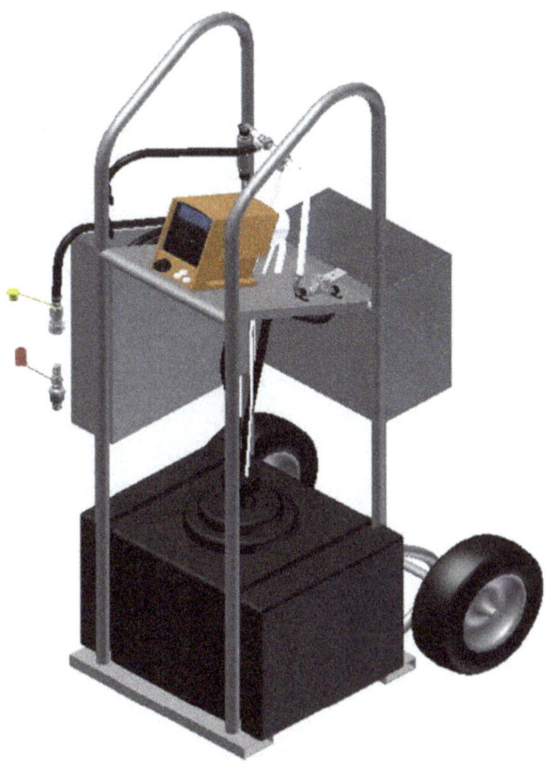

FIGURE 7.1. Mobile acid diffuser cleaning cart. Courtesy of Environmental Dynamics, International.

To assess the condition of diffusers, operators can resort to removal of diffusers for testing of OTEs in the laboratory. First manufacturers offer digital monitoring systems to monitor the condition of diffusers while tanks are in service.

8.0 MECHANICAL AERATION CONTROL

Mechanical aerators dissolve oxygen by thrashing the water surface to increase the surface area of water exposed to air. Means of adjusting oxygenation capacity to match the oxygen demand include submergence adjustment, speed adjustment, and on–off operation.

8.1 On-Time Adjustment

On–off switching of mechanical aerators may be the most practical way to adjust oxygenation capacity to oxygen demand, but it has pitfalls. Switching must not be so frequent that it would result in overload to the electrical equipment or controls. Nor should the aerator remain off for too long. Switching of multiple aerators should be desynchronized to avoid peak demand by having all mixers on at the same time.

Excessive off time may jeopardize the activated sludge process. A level of dissolved oxygen in the mixed liquor significantly below the target level may stress the microorganisms, or MLSS may settle.

8.2 Mechanical Aerator Maintenance a nd Troubleshooting

Surface or structural damage to the impellers or shrouds will impair the original energy efficiency. Prompt identification and repair of damage and removal of entrapped material such as rope will help maintain energy efficiency.

Because some mechanical aerators are inseparable from their motors, scheduled lubrication and preventive maintenance of motors are essential. Regular maintenance according to manufacturer's recommendations is critical to energy efficiency and reliable operation. Personnel should note noise, vibration, and temperature routinely and conduct periodic electrical and mechanical checks.

Motors, especially those on floating aerators, can become coated with aerosols of the mixed liquor that dry out and form a surface coating. If the coating becomes too thick, it will act as an insulator on the motor's cooling fins and will lead to motor burnout if not removed routinely. Routine hosing of the motors should be a preventive maintenance activity and, as such, included in the preventive maintenance program. Icing is another concern in some colder areas of the country. To prevent straining supports or overloading motors, ice should not be allowed to accumulate.

9.0 PROCESS MONITORING AND INSTRUMENTATION

This section covers aeration system monitoring needs to ensure energy-efficient operation. Probe and analyzer technologies and maintenance needs are covered in Chapter 9.

9.1 Dissolved Oxygen Measurement

Measurement of dissolved oxygen within biological reactors provides reference readings for both manual and automated control for both diffused and mechanically aerated systems. Target dissolved oxygen concentrations on operational considerations. In plug-flow reactors, dissolved oxygen concentrations can vary significantly along the flowpath of the tank. A single dissolved oxygen probe located in the front, middle, or end of the tank can be insufficient to adequate control aeration.

In coordination with dissolved oxygen monitoring, design provisions need to include means for operations to control airflow to the respective process area. Automated air control valves for diffuser grids will allow more accurate control than valves that require operations to make manual adjustments throughout the day. Supervisory control and data acquisition (SCADA) integration of any probes' readings used for process control is necessary so that operations staff can monitoring trends and assess meaningful cause–effect relationships for process optimization. Poorly controlled aeration systems will result in a high variability of actual dissolved oxygen readings compared to the target dissolved oxygen concentration. Covariances (defined as the standard deviation of 5 to 15 min dissolved oxygen data over a minimum of a day divided by the average actual dissolved oxygen concentration) of more than 25% are common for manually controlled aeration basins. Well-tuned proportional-integral-derivative (PID) based dissolved oxygen control or ammonia-based aeration control can reduce covariances to 10% to 20%. Model predictive control approaches are able to further reduce dissolved oxygen variability measured as covariance below 10%.

9.2 Oxidation Reduction Potential Measurement

Placement and use of an oxidation reduction potential (ORP) probe is typically within an anaerobic or anoxic zone to determine the presence or lack of inorganic electron acceptors (NO_3 and O_2). Oxidation reduction potential monitoring can serve here as a qualitative indicator of oxygen poisoning. This can occur when recycle streams carry back dissolved oxygen, sometimes at such low concentration that it is not quantified with dissolved oxygen probes. Increasing ORP values suggest that a reduction of aeration at the end of the aeration basins can improve BNR performance and reduce energy consumption. Operational adjustments can be made to RAS rate, internal mixed liquor recycle (IMLR) rate, and dissolved oxygen setpoints based on the relative ORP reading for the respective zone. In some cases, ORP probes have been used for automated control of the IMLR rate.

In systems operating at low dissolved oxygen concentrations, such as oxidation ditches or SNDN reactors, ORP probes may provide more stable readings than dissolved oxygen probes. In case dissolved oxygen concentrations monitored vary too rapidly over short periods of time, for example because of varying flow patterns in oxidation ditches, this can result in oscillations and dissolved oxygen control instabilities. Some utilities have reported more stable control and blower operation when using ORP instead of dissolved oxygen as the control input.

Typically, measurements received from an ORP probe are relied on as qualitative measurements to indicate the environment and not a reliable quantitive measurement.

9.3 Ammonium and Nitrate Measurement

Implementation of either ammonium or nitrate measurement can be for process monitoring or as a means of control. Ammonium oxidation can often be used as an indicator of relative process performance—above target ammonium concentrations can indicate insufficient oxygen, high loading events, inadequate SRT, or process upsets. The primary focus of online ammonium measurement as it relates to energy within water resource recovery is on target dissolved oxygen concentrations to reduce aeration energy inputs, supplying only enough aeration to meet process demands by altering dissolved oxygen setpoints and thus ammonia removal profiles. Online nitrate measurements focus on optimization of IMLR rates to meet effluent requirements and provide beneficial carbon oxidation. Ammonia and nitrate online readings have also in recent years been used as control inputs for aeration control systems that target total inorganic nitrogen setpoints.

Depending on the application, an in situ probe can be selected for more instantaneous feedback and faster trending of measurement. If accuracy and a high level of precision is required, potentially for stringent effluent permits, an analyzer is often selected. For implementation of either technology, site-specific testing of manufacturer probes/analyzers is recommended to determine the best alternative.

9.4 Airflow Measurement

Airflow measurements are used for operational control of aerated biological basins. Airflow demand can serve as a direct setpoint to blower control (airflow control method), allowing blowers to adjust pressure and airflow. Measurements are used in an automated control strategy that adjusts valves and blower operating points to meet target airflows, or dissolved oxygen, ammonia or nitrate target concentrations. Airflows per diffuser provide information about the OTE and can help operations staff to balance demand across diffuser grids to optimize OTE.

Airflow meters need to be properly sized for minimum and maximum airflows. Their accuracy can be verified through aeration system-wide airflow balances. Airflow comparisons between blower header and cumulative flows through air drop legs can help identify and quantify air leaks in damaged above or underground air yard piping, which can be substantial some facilities. For energy-efficient operation, it is recommended to place airflow meters on the blower header as well as on each diffuser grid drop leg equipped with manual or automated control valves.

9.5 Pressure Measurement

Pressure measurement is typically used on the air supply lines to aeration basins, on both individual zone drop legs and main supply lines. At times, flowmeters cannot be installed, for example, because of costs or available space. As an alternative, pressure measurement can be incorporated into an automated control scheme (such as with an orifice airflow meter) to control the air supplied. Herein, air valve positions in conjunction with pressure meters can be calibrated to predict the airflow through the valve. Pressure indicators can indicate if adequate pressure is supplied to the system. Inadequate pressure, below diffuser manufacturer recommendations, can result in decreased performance with unbalanced air distribution.

Another application is the use of pressure indicators on air supply headers; this can be incorporated into automated blower controls. Often blower controls are based on constant pressure setpoints in the combined blower discharge air header. Dynamic pressure control can save energy compared to constant pressure control, while airflow control can be the most energy-efficient control method.

9.6 Aeration Control

Because of the instantaneous variations in flow, load, and process conditions throughout the day, automated control of airflow variation to match the current oxygen demand will result in significant power savings to the WRRF.

Aeration control approaches used by facilities range from basic "set it and forget it" to manual adjustments and proportional-integral-derivative (PID) control (see Chapter 9 for further discussion on control and refer to previous section on how control inputs are measured). Basic PID control approaches use dissolved oxygen, ORP, or airflows as inputs. Advanced PID control approaches include ammonia-based aeration control or ammonia and nitrate-based aeration control systems. Advanced aeration control approaches that have been introduced into the wastewater industry over recent years use model predictive control systems, which use statistical methods to predict required control adjustments in advance based on process data patterns. These various methods differ in complexity, control stability and reliability, and their energy saving potential.

10.0 ENERGY-EFFICIENT AERATION SYSTEM PROCUREMENT

Owners, designers, and regulators can set clear expectations for energy efficiency during the design and procurement process of new aeration systems or system components.

Design specifications should be developed based on a comprehensive analysis of the complete aeration system even if only a single equipment item is procured. Some aeration systems such as surface aerators are self-contained. Other aeration systems will consist of a number of equipment units such as blowers, aeration piping, airflow control valves, diffusers, on-line instrumentation, and a control system that need to work seamlessly together.

The design of any mechanical aeration component needs to consider current and future process conditions, operational limits of equipment, and minimum mixing requirements as described in this chapter. In situ offgas testing and condition assessments of existing diffusers are a cost-effective option to assess the remaining life of existing assets before making replacement decisions.

Unless site conditions warrant simple manual aeration control (e.g., small package plants with marginal power demand), all new aeration systems and significant modifications to existing systems should evaluate the type of aeration control system is the best fit for the specific WRRF.

Owners are advised to request that designers include process model calibration and simulations as one component for making aeration system design decision. Calibrated and validated process modeling can provide an accurate analysis of the variations in aeration airflow requirements during different operational conditions, providing information for development of flexible, robust aeration and automated control systems. Results from the process model also provide the maximum, minimum, and typical airflow required during startup and future design conditions for correct sizing of the aeration blowers, air piping, airflow control valves, diffuser grids, supplemental mixers, and baffle walls within the aeration basin. Some state regulatory agencies have started to accept process model simulations, other advanced spreadsheet calculations, and historical facility performance data in the justifications of variances from aeration system design criteria that are empirical and conservative.

Specifications should explicitly request submission of information and performance requirements or guarantees that will allow owners and designers to assess the lifetime energy requirements for proposed equipment to compare different vendor products. Energy-efficiency guarantees need to be verified during shop and field testing of equipment. Owners should further request that training provided by vendors during commissioning explicitly addresses energy-efficient operation practices. Finally, facility staff is encouraged to document such practices in operation and maintenance manuals, standard operating procedures, and SCADA visuals used on a day-to-day basis for operational decision making.

11.0 AERATION AND ENERGY OPTIMIZATION

Energy optimization of an aeration system can have the most significant effect on a facility's total energy usage. Several opportunities exist within an aeration system that include both design and operational considerations.

11.1 Undesired Oxygen Carry-Over

Recycle streams such as mixed liquor or RAS recycle can carry residual dissolved oxygen concentrations that can "poison" anaerobic or anoxic zones. Energy relevant effects of dissolved oxygen carry-over is over-aeration at the end of the aeration basins and the oxidation of organic carbon that could be better used for nutrient removal or energy recovery.

11.1.1 Recycle Streams

Recycle streams can include internal (in-basin recycle) and external (RAS from clarification and solids dewatering/thickening streams). These streams can recycle excessive concentrations of dissolved oxygen and nitrogen oxide (NO_x) that, depending on the specific zone configuration, can be detrimental to process performance and system energy efficiency.

Excessive RAS recycle rates can re-introduce undesirable NO_x into initial anaerobic selector zones. Return activated sludge recycle rates typically range from 50% to 125% of influent forward flows and should be adjusted based on site-specific experience. Return activated sludge recycle flow can be mixed with primary effluent and directed to initial biological basin zones. Alternative configurations include RAS recycle into anoxic zones for RAS denitrification; these zones can be incorporated within the basin configuration or occur as separate process tanks. Return activated sludge denitrification mitigates contamination of anaerobic zones with NO_x components that can reduce process efficiency and biological selection.

Internal mixed liquor recycles provide environmental conditions for nitrification and denitrification. Nitrified IMLRs containing NO_x are recycled at the end of an aerobic zone back to an initial anoxic selector zone that provides carbon from primary effluent for denitrification. Another example is denitrified IMLR, returning MLSS from an anoxic zone into an anaerobic zone (similar to RAS denitrification). Internal mixed liquor recycle rates typically range from 2 to 3x the influent forward flow and can be adjusted to meet process requirements.

Both types of recycle, internal and external, can be introduced into the selected zones via mixing chimneys to mitigate the effect of dissolved oxygen carry-over. Adequate mixing of recycling streams with forward flow of upstream zones prevents short-circuiting and maintains desired conditions of the respective zone. However, if carry-over dissolved oxygen or NO_x is at detrimental levels, other measures can be taken to mitigate the effect.

Monitoring and control of recycle streams with the use of probes and analyzers provides indicators of zone conditions. Monitoring parameters include ORP, nitrate, dissolved oxygen, and flowrate. The combination of these probes can indicate if adjustments are required to recycle flowrates based on the ORP (anoxic vs. anaerobic) desired for each of the zones. Further nitrate measurements indicate the level of denitrification occurring within an anoxic zone and can be used to further refine recycle flows for optimal performance. Within aerobic zones, one measurement of dissolved oxygen at the recycle location can be managed to reduce the concentration of oxygen in the recycle.

11.1.2 Baffle Wall Design

Undesired dissolved oxygen carry-over can be mitigated by process control of recycle streams as well as physical barriers. Baffle walls can be implemented as physical barriers to separate zones of different environmental conditions. Baffle walls are commonly implemented in nutrient removal applications in which separation of oxic, anoxic, and anaerobic zones is critical to process efficiency. In addition to traditional anaerobic and anoxic zone separation, zones of significantly differing oxic conditions can consider the use of baffle walls to prevent back and forward mixing dissolved oxygen concentrations. Flow into subsequent downstream zones can be managed with several baffle wall designs.

- Overflow Weir Baffle—Baffle walls with overflow weirs allowing the free flow of filaments and scum forward into downstream zones.

- Over-Under Baffle—Baffle walls incorporating a combination of overflow weirs and underflow openings (located at base of baffle wall). These allow passing of filaments and scum from zone to zone while also limiting dissolved oxygen carry-over. The underflow openings limit air entrainment that can occur with overflow weirs.
- Ported Baffle—Baffle walls allow flow through multiple ported openings within the baffle. This limits air entrainment but does not provide a solid barrier protecting against back mixing.
- Serpentine—Baffle walls that force flow linearly from one zone to the next. Flow passes between zones via openings located at opposite corners of the basin creating a serpentine flow pattern.

Consideration of baffle technology and implementation includes:

- Air Entrainment—Depending on specific application, the mitigation of air entrainment can protect anaerobic zones. Weir designs should limit freefall and turbulence.
- Short-Circuiting—Zone hydraulic conditions can be evaluated with CFD modeling to determine mixer placement and zone design elements. For zones where two or more influent streams are entering, the use of a mixing chamber can prevent short-circuiting and efficiently mix the separate streams.
- Plugging—Potential for rag buildup and grit/solids accumulation can limit the effectiveness of technologies such as underflow and ported baffles. Hydraulic velocities and dynamics can create conditions to mitigate buildup.
- Surface Solids Accumulation—Accumulated scum and surface solids must be allowed to pass from zone to zone.

11.2 Aeration Energy Monitoring and Benchmarking

11.2.1 Typical Aeration Energy Demand

Of wastewater treatment energy costs, aeration accounts for 50% to 75% of total energy in WRRFs (Gao et al., 2014; Gori et al., 2011; Jiang & Stenstrom, 2012; Wang et al., 2013; Zhou et al., 2013). A state-of-the-art review for energy benchmarking throughout the world was performed by Longo et al. (2016) (see also Appendix B). This review included multiple metrics of measurement including energy usage based on volume of water treated, population equivalents, and chemical oxygen demand removed. Biological nutrient removal systems, compared to conventional activated sludge systems, show the potential to implement more efficient processes while also treating to a higher effluent standard. This difference is attributed to process optimization that includes automation and implementation of nutrient removal (beneficial carbon oxidation via denitrification). Aeration energy from the study was found to range from 0.18 to 0.8 kWh/m^3 (680 to 3000 kWh/mil. gal). Aeration energy should be benchmarked on multiple metrics to allow for comparison to similar facilities, however, at minimum it is recommended to use the kWh/volume (i.e., gallon or liters) International Standard. Refer to Chapter 2 for a detailed discussion on benchmarking.

11.2.2 Meaningful Supervisory Control and Data Acquisition Screens

Operations staff seldom have routine access to energy consumption records of their facilities, which are often handled by separate accounting departments. Nor is real-time power consumption data available. Therefore, energy consumption of process operation is not a decision factor in daily operational decision making. Some first facilities have started to integrate real-time power monitoring data into SCADA systems so that operations staff can consider this information for overall process control and optimization. When energy consumption data is integrated into SCADA systems, it is recommended to develop meaningful criteria that facilitate process control decisions. Examples include aeration power per BOD or ammonia load removed, or a visual alarm of power demand threshold exceedances that trigger peak demand charge increases.

12.0 ECONOMIC IMPLICATIONS

Investment into energy-efficient aeration process selection, equipment, control systems, and operational practices have resulted in significant capital and operational cost savings for utilities in the United States. Success stories are broadly published (U.S. EPA, 2010). Economic benefits often are accompanied by other benefits, such as reduced risk of permit violations, improved effluent quality, reduced wear and tear on equipment, and increased employee satisfaction and retention.

13.0 REFERENCES

Barber, W., Schroedel, R., Shimada, T., Spalding, S., Whitlock, D., & Williams, M. (2019). *High performance anaerobic digestion* [Fact sheet]. Water Environment Federation. https://www.wef.org/globalassets/assets-wef/3---resources/topics/a-n/biosolids/technical-resources/high-performance-ad-final.pdf

Barillon B., Ruel, S. M., Langlais, C., & Lazarova, V. (2012). Energy optimization in membrane bioreactors. In: *Proceedings of IWA World Water Congress & Exhibition 2012*, 16–21 September, Busan, Korea.

Bowden, G., Tsuchihashi, R., & Stense, H. D. (2016). Technologies for side-stream nitrogen removal. *Water Intelligence Online*, *15*. 9781780407890–9781780407890. https://doi.org/10.2166/9781780407890

Daigger, G. T., & Bailey, E. (2000). Improving aerobic digestion by prethickening, staged operation, and aerobic-anoxic operation: Four full-scale demonstrations. *Water Environment Research, 72*(3), 260–270. https://doi.org/10.2175/106143000X137464

Downing, A. L., & Scragg, L. J. (1958). The effect of synthetic detergents on the rate of aeration in diffused-air activated sludge plants. *Water & Waste Treatment Journal*, 7(3), 102–107.

Fitzgerald, C. M., Camejo, P., Oshlag, J. Z., & Noguera, D. R. (2015). Ammonia-Oxidizing microbial communities in reactors with efficient nitrification at low-dissolved oxygen. *Water Research, 70*, 38–51.

Gao, H., Scherson, Y. D., & Wells, G. F. (2014). Towards energy neutral wastewater treatment: Methodology and state of the art. *Environmental Science: Processes Impacts, 16,* 1223–1246.

Gori, R., Jiang, L.-M., Sobhani, R., & Rosso, D. (2011). Effects of soluble and particulate substrate on the carbon and energy footprint of wastewater treatment processes. *Water Research, 45*, 5858–5872.

Hanaki, K., Wantawin, C., & Ohgaki, S. (1990). Nitrification at low levels of dissolved oxygen with and without organic loading in a suspended-growth reactor. *Water Research, 24*(3), 297–302.

IWA. (2018). In D. Ross (Ed.), *Aeration, mixing, and energy—Bubbles and sparks*. IWA Publishing.

Jenkins, T. E. (Ed.). (2013). Aeration processes. In *Aeration control system design* (pp. 49–76). John Wiley & Sons, Inc. https://doi.org/10.1002/9781118777732.ch03

Jiang, P., & Stenstrom, M. (2012). Oxygen transfer parameter estimation: Impact of methodology. *Journal of Environmental Engineering—ASCE, 138*(2), 137–142.

Johnson, B. R., Menniti, A., & Murthy, S. (2011). Impacts of post aerobic digestion on the design of nutrient removal facilities. *Proceedings of the Water Environment Federation, 2011*(1), 503–509. https://doi.org/10.2175/193864711802867559

Keene, N., Reusser, S. R., Scarborough, M., & Noguera, D. R. (2017). Pilot plant demonstration of stable and efficient high rate biological nutrient removal with low dissolved oxygen conditions. *Water Research, 121*, 72–85.

Longo, S., D'Antoni, B. M., Bongards, M., Chaparro, A., Cronrath, A., Fatone, F., Lema, J. M., Mauricio-Iglesias, M., Soares, A., & Hospido, A. (2016). Monitoring and diagnosis of energy consumption in wastewater treatment plants. A state of the art and proposals for improvement. *Applied Energy, 179*(1) 1251–1268.

Metcalf & Eddy. (2014). *Wastewater engineering: Treatment and resource recovery* (5th ed.).

Tchobanoglous, G., Burton, F. L., & Stensel, H. D. (2002). *Wastewater engineering: Treatment and reuse* (4th ed.). McGraw-Hill Science/Engineering/Math.

Turk, O., & Mavinic, D. S. (1987). Benefits of using selective inhibition to remove nitrogen from highly nitrogenous wastes. *Environmental Technology Letters, 8*(1–12), 419–246. https://doi.org/10.1080/09593338709384500

U.S. Environmental Protection Agency. (1989). *Design manual: Fine pore aeration systems.*

U.S. Environmental Protection Agency (2010, September). *Evaluation of energy conservation measures for wastewater treatment facilities* (EPA 832-R-10-005.3).

van der Star, W. R. L., Abma, W. R., Blommers, D., T., Mulder, J.-W., Tokutomi, T., Strous, M., Picioreanu, C., & van Loosdrecht, M. C. M. (2007). Startup of reactors for anoxic ammonium oxidation: Experiences from the first full-scale anammox reactor in Rotterdam. *Water Research, 41*(18), 4149–4163. https://doi.org/10.1016/j.watres.2007.03.044

Wang, X., Liu, J., Qu, B., Ren, N. Q., & Qu, J. (2013). Role of carbon substrates in facilitating energy reduction and resource recovery in a traditional activated sludge process: Investigation from a biokinetics modeling perspective. *Bioresource Technology, 140*, 312–318.

Water Environment Federation. (2017). EPA ECHO Biosolids data summary.

Water Environment Federation. (2018). *Design of municipal wastewater treatment plants* (5th ed.; WEF Manual of Practice 8; ASCE Manual of Practice No. 76). McGraw-Hill: McGraw-Hill Education.

Wett, B., Nyhuis, G., Takács, I., & Murthy, S. (2010). Development of enhanced deammonification selector. *Proceedings of the Water Environment Federation, 2010*(10), 5917–5926. https://doi.org/10.2175/193864710798194139

Zhou, X., Wu, Y., Shi, H., & Song, Y. (2013). Evaluation of oxygen transfer parameters of fine bubble aeration system in plug flow aeration tank of wastewater treatment plant. *Journal of Environmental Sciences, 25*(2), 295–301.

14.0 SUGGESTED READINGS

Casey, E., Glennon, B., & Hamer, G. (1999). Review of membrane aerated biofilm reactors. *Resources, Conservation and Recycling,* 27(1), 203–215. https://doi.org/10.1016/S0921-3449(99)00007-5

Leu, S.-Y., Chan, L., & Stenstrom, M. K. (2012). Towards long solids retention time of activated sludge systems: Benefits in energy savings, effluent quality, and stability. *Water Environment Research, 84*(1), 42–53.

Rosso, D., & Stenstrom, M. K. (2006). Economic implications of fine-pore diffuser aging. *Water Environment Research, 78*(8), 810–815.

Rosso, D., & Stenstrom, M. K. (2017). Energy savings from denitrification. *Environmental Engineer: Applied Research and Practice,* Vol. 3, Summer 2017.

Rosso, D., & Stenstrom, M. K. (2005). Fifteen years of offgas oxygen transfer efficiency measurement on fine-pore aerators: Key role of sludge age and normalized air flux. *Water Environment Research, 77*(3), 266–273.

8

Blowers

Thomas E. Jenkins, PE; Mark K. Addison, PE; & William Thies, PE

1.0 INTRODUCTION

Proper consideration of air requirements and blower energy consumption is an essential part of both facility design and energy conservation in existing water resource recovery facilities. Aeration blowers typically consume more than 50% of the total energy in diffused aeration activated sludge secondary treatment facilities (see Chapters 4 and 7). Secondary aeration is the focus of this chapter, but blowers are also used in many other processes, including grit removal, channel aeration, aerobic digestion, and post-aeration.

As important as blowers are to energy consumption, they are even more important to maintaining proper process performance. An adequate supply of air to the treatment process is essential to meet permit limits. Process compliance considerations should, in all cases, override energy conservation.

Blowers may be used for compressing gases other than air, such as methane for mixing anaerobic digesters (see Chapter 10). Many of the same types of equipment and application considerations apply for any gas being compressed. The blower manufacturer should be consulted for performance evaluations when compressing gases other than air.

2.0 APPLICATION CONSIDERATIONS

There are many similarities between pumps (see Chapter 5) and blowers. Positive-displacement and centrifugal designs are available for both. The energy consumption of both pumps and blowers is a function of flowrate, discharge pressure, and equipment efficiency. A variety of control methods can be used to control either pumps or blowers. However, the compressibility of air makes blower application and control more complex than pump application.

A blower manufacturer's design and selection procedures include rigorous calculations and consideration of details that may not be significant in routine energy evaluations. The simplified formulas presented herein are sufficiently accurate for comparisons of various systems and verification of the cost-effectiveness of conservation measures. The blower manufacturer should be consulted for final confirmation of all designs.

2.1 Effects of Compressibility

Air is a mixture of gases. Dry air is predominantly nitrogen (75.5%, by weight) and oxygen (23.1%, by weight), with the balance composed principally of argon and carbon dioxide. Atmospheric air also contains water vapor, with the amount varying from negligible to several percent by weight. The air temperature and relative humidity define actual moisture content.

The density of any fluid varies with temperature and pressure. With liquids, this variation is typically negligible. For gases such as air, however, the variations in density with temperature and pressure are so pronounced that they must be considered even in ordinary applications. Therefore, when dealing with airflow, the thermodynamic state of the air stream under question must always be defined.

In the United States, "standard" air conditions for wastewater treatment applications are 14.7 psia, 68 °F, and 36% relative humidity (101.3 kPa, 20 °C, and 36% relative humidity). A cubic foot of air at these conditions has a mass of 0.034 kg (0.075 lb). Airflow rates are often expressed as standard cubic meters per second (standard cubic feet per minute [SCFM]). Although the units represent a volumetric flowrate, because of the defined properties and density, SCFM also identifies the mass flowrate of air and oxygen.

In the United States, volumetric flowrate is generally expressed as actual cubic feet per minute (ACFM) or inlet cubic feet per minute (ICFM), where ICFM refers to ACFM at the inlet to the blower. The relationship between SCFM and the ACFM required to deliver an equal mass flowrate of oxygen is

$$\text{ACFM} = \text{SCFM} \cdot \frac{14.58}{p_b - (RH \cdot p_{va})} \cdot \frac{460 + T_a}{528} \cdot \frac{p_b}{p_a} \tag{8.1}$$

Where

RH = ambient air relative humidity, decimal;
p_{va} = saturated vapor pressure of water at actual temperature, psia;
T_a = actual air temperature, °F;
p_a = actual air pressure, psia; and
p_b = barometric pressure, psia.

Outside of the United States, volumetric flowrate is generally expressed as m^3/h, and "standard" airflow rate is often specified as normalized cubic meters per hour (Nm^3/h). The specification varies, but 101.3 kPa, 0 °C, and 0% relative humidity (14.7 psia, 32 °F, and 0% relative humidity) are common. A normalized cubic meter of air at these conditions weighs 1.293 kg (2.85 lb).

$$\frac{\text{m}^3}{\text{hr}} = \frac{\text{Nm}^3}{\text{hr}} \cdot \frac{101.3}{p_b - (RH \cdot p_{va})} \cdot \frac{273 + T_a}{273} \cdot \frac{p_p}{p_a} \tag{8.2}$$

Where

RH = ambient air relative humidity, decimal;
p_{va} = saturated vapor pressure of water at actual temperature, kPa;
T_a = actual air temperature, °C;
p_a = actual air pressure, absolute kPa; and
p_b = barometric pressure, absolute kPa.

For preliminary calculations, relative humidity can be ignored, and the above equations may be simplified to Boyle's Law and Charles's Law. To convert airflow rate from one set of conditions to another use the following equation:

$$q_{v2} = q_{v1} \cdot \frac{p_1}{p_2} \cdot \frac{T_1}{T_2} \quad (8.3)$$

Where

q_v = flowrate in volumetric units, m^3/h or Nm^3/h (ACFM or SCFM);
p = absolute pressure, absolute kPa (psia); and
T = absolute temperature, K (°R).

2.2 Common Blower Types

The most common blowers used in wastewater treatment can be divided into two general categories: *positive-displacement* and *centrifugal.*

Positive-displacement (PD) blowers provide a fixed volume of air at any given speed, providing a constant volumetric flowrate (inlet m^3/h or ICFM) at variable discharge pressure. There are two principal designs of PD blowers: the lobe type, with airflow transverse to the blower shaft, and the screw type, with airflow parallel to the blower shaft.

Centrifugal blowers are often referred to as "dynamic" machines. They consist of one or more impellers on a rotating shaft. The impellers use vanes like those on a centrifugal pump impeller to transfer kinetic energy to the air. The diffuser section and volute of the blower convert the kinetic energy (velocity pressure) to potential energy (static pressure). In contrast to PD blowers, at fixed speed centrifugal blowers generally provide a wide range of flowrates over a narrow range of discharge pressures. Centrifugal blowers are further divided into single-stage and multistage types. Single-stage blowers may be geared or direct drive.

There are other types of blowers that are occasionally used in wastewater treatment applications, including liquid-ring, vane, and regenerative (side channel) blowers. These are generally used in low horsepower applications.

Historically, blowers were typically supplied as open skid-mounted units, with motors and some accessories mounted on the skid. Controls, drives, and many accessories were field installed. It is now common to have the blower, controls, and all necessary accessories factory assembled in a sound-absorbing enclosure. All types of blowers are commercially available in this type of factory package.

2.3 Blower Power Requirements

A useful term, which appears in many blower calculations, is "X", sometimes referred to as the *adiabatic factor*.

$$X = \frac{p_d}{p_i}^{\frac{k-1}{k}} - 1 \qquad (8.4)$$

Where

p_d = absolute discharge pressure, absolute kPa (psia);
p_i = absolute inlet pressure, absolute kPa (psia); and
k = C_p/C_v, dimensionless $k \approx 1.395$ for air.
Note: $k - 1/k \approx 0.283$ for air

Equation 8.4 assumes ideal gas behavior, and therefore the ratio of heat capacities is constant. The value of k actually varies slightly with temperature and relative humidity.

Another useful term used in many calculations is blower head. This is the work per unit mass done by the blower, expressed in J/kg (ft·lb_f/lb_m). Unlike pump head, which is identical to discharge pressure, the compressibility of air results in blower head being a function of several terms.

$$W = K \cdot T_i \cdot X \qquad (8.5)$$

Where

W = head or work, J/kg (ft·lb_f/lb_m);
T_i = inlet temperature, K (°R); and
K = conversion factor
1014 (S.I./metric)
188.5 (U.S. customary).

Head is useful in calculations for both PD and centrifugal blowers, but it is less often referenced in analyses of PD blowers. This is because the pressure ratio for PD blowers varies with system characteristics and is independent of airflow.

The power required for compressing air is a function of the flowrate and pressure ratio from inlet to discharge of the blower, as follows:

$$p = \frac{\rho \cdot q_i \cdot w}{\eta} = K \cdot \frac{q_i \cdot p_i \cdot w}{\eta} \quad (8.6)$$

Where

P = power, kW (horsepower);
W = head or work, J/kg (ft·lb$_f$/lb$_m$);
q_i = blower inlet volumetric flowrate, m^3/h (ICFM);
ρ = density of gas at inlet conditions, kg/m^3 (lb/cu ft);
p_i = absolute blower inlet pressure, kPa (psia);
η = efficiency, decimal;
X = blower adiabatic factor, dimensionless; and
K = conversion factor
9.816×10^{-4} (S.I./metric)
0.01542 (U.S. customary).

If the efficiency used in Equation 8.6 is for the bare blower, the calculated power is the power demand at the blower shaft. Additional energy losses for motors, drives, and other accessories should be added to the shaft horsepower to determine the total electrical power required, often referred to as "wire-to-air" power. If wire-to-air power is required, the wire-to-air efficiency must be used.

Blower efficiency is not constant. Efficiency for a specific blower varies with flowrate, pressure, speed, and inlet conditions. In general, larger blowers are more efficient than smaller blowers of the same design. Blower efficiency decreases at airflows greater than or less than the best efficiency point (BEP). The efficiency and power demand changes with inlet temperature, pressure, and relative humidity. The control method has a significant effect on power demand.

A common parameter for comparing blowers is the specific power, typically expressed as kW/m^3·s (kW/100 CFM). This is the total wire-to-air power required to deliver a given inlet volumetric flowrate at a given discharge pressure. It should be noted that the specific power reported is based on a unique set of conditions. If the discharge or inlet conditions change, the specific power must be calculated to reflect the new conditions.

2.4 Blower and System Curves

Blower specifications typically identify a single flowrate and discharge pressure at one inlet temperature, humidity, and pressure as the design point. This typically represents the worst-case operating conditions expected. Blowers can operate at a variety of flowrates and discharge pressures. The variation of discharge pressure plotted against flowrate represents the blower's "characteristic curve," which is often referred to as the *blower curve* (see Figure 8.1). Blower curves are specific to a single inlet temperature, pressure, relative humidity, and operating speed. They may use volumetric or standard condition airflow for the x-axis. The shape and slope of the curve varies with the type of vanes on the blower impeller(s).

The blower power consumption is typically plotted against flowrate in a separate curve, referred to as the *power curve* (see Figure 8.1).

The blower "curve" for a PD blower operating at constant speed approximates a vertical line. The curve for a centrifugal blower at constant speed is generally downward sloping, with the maximum discharge pressure to the left of the normal operating range.

A blower curve by itself does not identify the actual operating airflow and discharge pressure. It represents a range of possible operating conditions. The actual airflow rate is a function of static pressure and velocity pressure. Note that blowers create airflow, and the system resistance to flow creates pressure.

In most wastewater treatment processes the air is provided to a point below a water surface. The depth of liquid above the top of the diffusers or air release point creates static pressure, which is typically the largest portion of the total blower discharge pressure.

Additional pressure is created by the system's resistance to blower airflow. The friction from air flowing through piping, valves, and diffusers creates pressure drop, which varies with air flowrate and control-valve position. The plot of this variation, combined with the static pressure, is called the system curve. The intersection of the system curve and blower curve identifies the actual flowrate at specific conditions (see Figure 8.1). Aeration systems are typically designed to allow 7 to 14 kPa (1 to 2 psi) total pressure drop at worst-case operating conditions, which includes allowances for diffusers and flow control valves.

Most aeration systems have multiple tanks operating in parallel and several grids of diffusers in each tank. Flow control valves are provided to control the distribution of the airflow provided by the blowers between the various locations (see Chapter 9). The proper adjustment of the flow control valves is important to optimize energy consumption of the blowers and aeration systems.

Optimum sizing of piping and related components such as filters, valves, and silencers will have a direct effect on blower power requirements. Undersized discharge piping increases friction and pressure loss, which results in increased energy requirements. Pipe and valves sized to the velocity ranges in Table 8.1 will provide good control and reasonable friction losses.

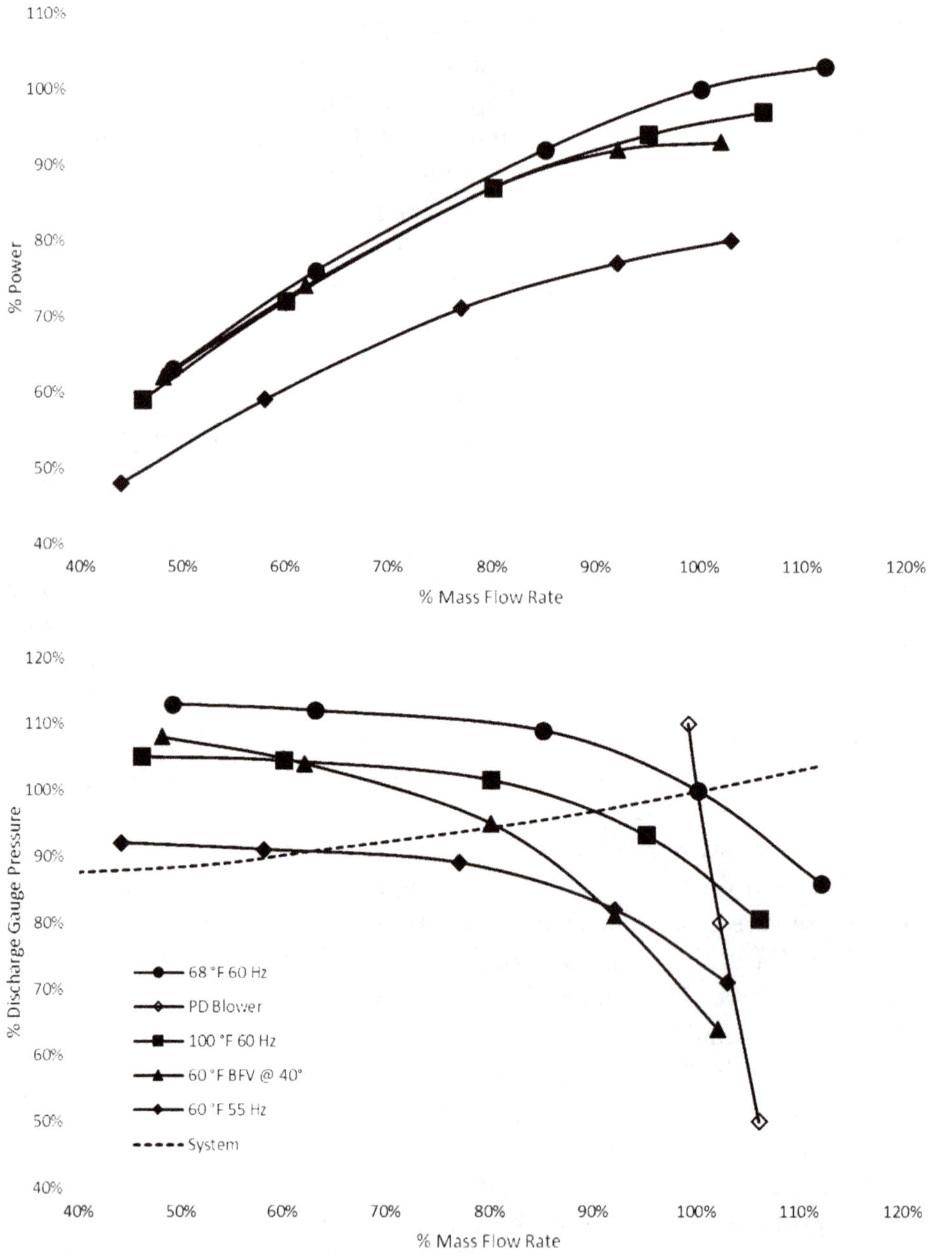

FIGURE 8.1. Example blower characteristic and system curves and power curves. Reprinted with permission from Howden USA.

Normal design practice is to have less than 3.5 kPa (0.5 psi) pressure drop between the ambient air intake point and the blower inlet at design airflow. Approximately half this allowance is for pressure drop through the inlet filter, with the balance for losses in inlet pipe and fittings. Inlet valves used for throttling centrifugal blowers and basin airflow control valves must

TABLE 8.1. Typical distribution piping air velocities at max airflow

Nominal Pipe Diameter	Design Velocity, feet per minute
1″ to 3″	1200 to 1800
4″ to 10″	1800 to 3000
12″ to 24″	2700 to 4000
30″ to 60″	3800 to 6500

be sized to provide adequate control while remaining within the useable range of valve travel, typically 15% to 70% open.

2.5 Effect of Inlet Conditions

Blower performance is greatly influenced by inlet conditions. A typical blower application encounters a wide range of temperature, pressure, and humidity. Increasing inlet temperature, decreasing inlet pressure, and/or increasing relative humidity decrease the density of the inlet air. A decrease in density increases the inlet volumetric flowrate the blower must provide for an equivalent mass flowrate, which has a direct effect on power requirements.

Inlet pressure varies with altitude, weather conditions, and inlet filter condition. The effect of altitude on barometric and inlet pressure can be significant. Barometric pressure decreases approximately 11 kPa/1000 m elevation (1 psi/2000 ft).

When evaluating energy requirements and comparing alternate system designs, it is important to make sure that all evaluations are made using the same assumptions for the inlet conditions. It is also important that the energy evaluation be made at typical operating conditions.

2.6 Other Considerations

Ambient conditions and their influence on blower power consumption are beyond the control and influence of the designer and operator. However, other design and maintenance procedures can improve energy efficiency.

An easily neglected operational consideration is proper adjustment of aeration basin airflow control valves to minimize system discharge pressure (see Chapter 9).

Blower power is proportional to airflow rate. It is critical to match blower airflow to process requirements (see Chapter 9). Some processes, such as channel aeration, filter backwash, and air lift pumps, depend on volumetric flowrate in m^3/h (ACFM) for proper performance. Other processes, such as secondary aeration and aerobic digestion, depend on the correct mass flowrate of oxygen to the process in Nm^3/h (SCFM).

Many processes, such as post-aeration or conventional diffused aeration, operate at fixed static pressure throughout the operating flow range. The only significant change in system pressure is the variation of friction pressure drop through air-distribution piping. Other processes, such as filter backwash, aerobic digesters, and sequencing batch reactors, show a wide variation in static pressure during normal operation. These processes can have changes of 50% or more in diffuser submergence, and the blower system must be able to accommodate variations in discharge pressure. With PD blowers, the discharge pressure inherently rises and falls to match system requirements. Centrifugal blowers, however, require automatic controls to maintain desired airflow during pressure variations.

Discharge temperature of air can be an important consideration for applications with deep aeration basins. Discharge temperature increases with higher inlet temperature, higher pressure ratio, lower speed (for lobe type PD blowers), and lower efficiency. High discharge temperature can cause damage to nonmetallic piping, gaskets, and seals. The highest discharge temperature should be calculated for each application to verify material suitability.

$$T_d = T_i + \frac{T_i \cdot X}{\eta_b} \tag{8.7}$$

Where

T_d = discharge temperature, K (°R);
T_i = inlet temperature, K (°R);
X = adiabatic factor, dimensionless (see Equation 8.4); and
η_b = efficiency of the bare blower, decimal.

3.0 POSITIVE-DISPLACEMENT BLOWERS

3.1 Common Operating Principles

Positive-displacement blowers at constant speed and inlet conditions operate at constant volumetric flowrate. In a PD blower, each revolution of the drive shaft causes the impeller lobes to sweep a fixed volume of air from inlet to discharge. Close clearances between the impellers and the inside of the case, and between the two impellers as they mesh, provide a seal that restricts the backward flow of air. Timing gears on each shaft maintain the clearances between the two lobes. The motor may be directly coupled to the blower shaft or use a V-belt drive or gears to operate the blower at a speed different from the motor speed. If the blower is belt-driven, additional power loss (typically 2% to 5%) will occur as the result of belt slippage.

The combined volume of air swept by the lobes during a full rotation is referred to as *blower displacement*, expressed in cubic meters per revolution (cubic feet per revolution).

In a PD blower, the discharge pressure inherently matches the pressure in the discharge piping. This makes application of PD blowers to variable-depth processes straightforward. Increasing discharge pressure increases the power demand of a PD blower.

Positive-displacement blowers are not as sensitive to changes in inlet conditions as centrifugal machines. A PD blower operating at constant speed will not exhibit any appreciable power change in response to an increase in inlet temperature, but the mass flowrate will decrease. A variable-speed PD blower experiencing an increase in inlet temperature will require increased speed and power to maintain a constant mass flowrate. If a blower's inlet pressure decreases and the discharge and barometric pressures remain constant, the pressure rise across the blower is increased. For a constant-speed PD blower, the result is a slight reduction in mass flowrate and an increase in power draw.

3.2 Control Techniques

The most practical control technique for PD blowers is varying blower speed. In small systems with belt drives and constant-speed motors, changing the blower flowrate to match average process demand may be accomplished by changing sheaves on the blower or motor. If the blower airflow is higher than process demand, changing sheaves to reduce blower speed will reduce flowrate and power requirements. If the discharge pressure is lower than design, or if the motor rating exceeds actual power draw, the blower speed can be increased to provide additional capacity. This may eliminate the need to run additional blowers at lower efficiency.

When changing the blower speed, it is important to stay within the manufacturer's speed limitations and motor power rating. If several blowers are available, it may be desirable to have different operating speeds for each of them. The operator should then select the blower(s) to run based on process demand.

In systems with automatic controls, the blower airflow should be continuously adjusted to exactly meet process requirements. Several devices are available for this, including variable-pitch sheaves, eddy current drives, and magnetic couplings. The most common device used is the variable frequency drive (VFD). Manual or simple automatic controls can be used with VFDs and PD blowers. Positive-displacement blowers are a constant torque load, and VFDs should be selected accordingly.

Positive-displacement blower power is composed of two parts: *gas horsepower* and *friction horsepower*. The gas horsepower represents the power to compress the air from inlet to discharge conditions. This is the largest portion of the total blower power. Additional power is required because of losses in bearings, gears, and seals. This is commonly referred to as *friction horsepower.*

Positive-displacement blower airflow delivery to the process may also be controlled by using unloading or blow-off valves to dump a portion of the airflow to the atmosphere. This technique does not reduce the power required by the blower and should be avoided. The pressure relief valve should never be adjusted to dump excess airflow. Positive-displacement blowers should never be throttled, as this induces additional load on the rotors and may lead to catastrophic failure. Control valves at aeration basins or other process locations should only be used to distribute air in proportion to the basin process requirements. Adjusting these valves will not change the blower airflow rate.

3.3 Lobe Type Positive-Displacement Blowers

In a lobe type PD blower, the impellers sweep the volume air around the outside of the case from inlet to discharge (see Figure 8.2). Compression takes place as the volume of air is pushed into the system. This is often called *external compression.*

Leakage of air from the discharge back to the inlet of the blower, caused by imperfect impeller sealing, is referred to as *slip* and reduces blower efficiency. To compensate for the slip, the blower must operate at a slightly higher speed so that the correct discharge flowrate will be produced.

3.4 Screw Type Positive-Displacement Blowers

A second type of PD blower is the rotary screw blower. It uses two counter-rotating helical impellers to move a volume of air from inlet to discharge. One impeller has female lobes and the other has male lobes. As the impellers rotate, the volume trapped between them moves from one end of the

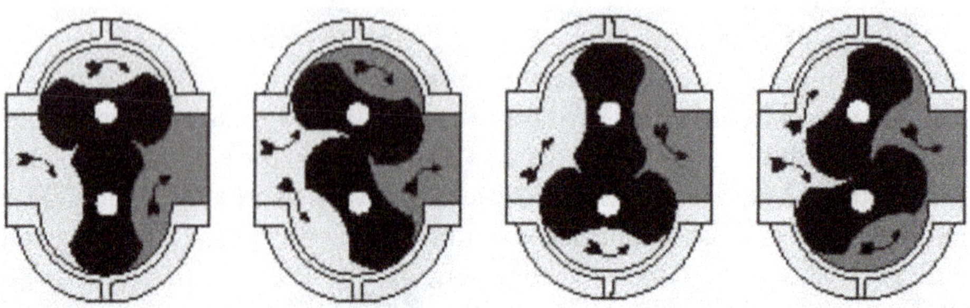

FIGURE 8.2. Lobe type PD blower. Reprinted with permission from Howden USA.

impeller to the other. The volume between the lobes is reduced at a fixed ratio, and pressure increases at a corresponding ratio. When the air reaches the end of the impeller, it is forced into the system and the pressure rises or falls to match system pressure (see Figure 8.3).

Screw blowers should not be confused with lobe blowers that use helical lobes to reduce noise and pulsation. In the lobe blower, the air moves around the outside of impeller transverse to the blower shafts. In the screw blower, the air moves axially.

Most of the screw blower's compression takes place internally between the rotors, which is efficient. However, if the system pressure does not match the pressure determined by the volume ratio of the impellers, losses result. If the system pressure is lower, the result is over-compression and wasted power. If the system pressure is higher, the result is under-compression, and the additional pressure increase takes place less efficiently external to the impellers. Under-compression is less stressful on the blower internal components. However, unless there is a significant mismatch between the internal pressure increase and the system pressure, screw blowers will be more efficient than lobe blowers.

Unlike lobe blowers and many centrifugal blowers, the peak efficiency of screw blowers occurs at approximately the middle of the operating range. In many applications, this improves the average energy consumption when evaluated across the blower duty cycle. Another feature of the screw blower is excellent turndown, which often exceeds 60%. This flexibility can be important for many facilities during periods of low loading.

3.5 Application Considerations

Precautions should be taken in system design to minimize the operator's exposure to noise. Inlet and discharge silencers are commonly used, and piping in the blower room may be insulated to reduce sound radiation. Acoustic enclosures are often provided to contain noise. Some lobe blower designs

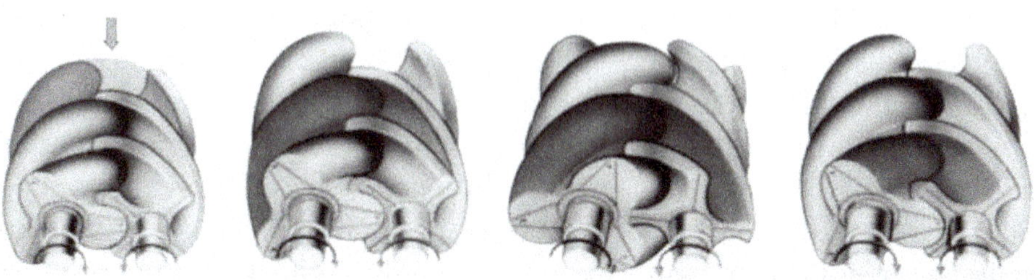

FIGURE 8.3. Screw type PD blower. Reprinted with permission from Aerzen USA Company, Coatesville, Pennsylvania.

use three lobed or twisted (helical) impellers to simplify noise attenuation, although these designs may reduce energy efficiency.

The discharge of PD blowers should always be equipped with a pressure-relief valve. This prevents damage to equipment or piping resulting from excessive pressure. Some systems use an unloading valve to reduce motor torque and current during starting.

3.6 Operating Limits

Excessive air temperature rise will change internal clearances and must be avoided. Temperature typically increases as speed decreases, and the blower should not be operated below the manufacturer's recommended minimum speed. Motor overheating may also result if speed is reduced to a point at which motor ventilation is compromised.

Blower efficiency for lobe units will decrease as speed is reduced, so it is tempting to operate at a high speed. However, noise and mechanical stress increase significantly as speed increases. Bearing lubrication and stress on rotating members determine the upper limit of blower speed. As the difference between inlet and discharge pressure increases, mechanical stresses increase. This may limit the allowable pressure differential across the blower.

4.0 CENTRIFUGAL (DYNAMIC) BLOWERS

4.1 Common Operating Principles

The various types of centrifugal blowers share many operating characteristics. Efficiency and head are constant at a given speed and inlet volumetric flowrate. Discharge pressure and power demand at a given airflow rate change with inlet pressure, temperature, and relative humidity. The blower and power curves shift with these changes (see Figure 8.1).

As the inlet pressure decreases, the discharge pressure and power decrease at a given volumetric inlet flowrate if all other inlet and operating conditions remain constant. If a blower's inlet pressure decreases and the discharge and barometric pressures remain constant, the pressure rise across the blower is increased. For a constant-speed centrifugal blower, the mass and volumetric flowrates are reduced, and blower power draw is decreased. For a variable-speed centrifugal blower with constant ambient conditions and mass flow requirements, pressure drop across the inlet because of inlet piping, filter losses, and other factors will require an increase in pressure ratio and inlet flow (to compensate for the reduced inlet density), resulting in increased power draw.

Inlet temperature is the most significant variable affecting centrifugal blower performance. It affects density and pressure ratio as well as power and discharge temperature. When a blower is modulated to provide a constant mass flowrate, an increase in inlet temperature will decrease the power required for an inlet-throttled centrifugal blower but will increase the power required for a variable-speed centrifugal blower.

The American Society of Heating, Refrigerating and Air-Conditioning Engineers and the National Oceanic and Atmospheric Administration publish data on temperature variations for the United States.

The effect of inlet conditions on the performance of various types of centrifugal blowers can be seen in Figure 8.1.

4.2 Control Techniques

The ability to vary airflow using a variety of devices is one of the advantages of centrifugal blowers. In some cases, blower airflow is controlled by discharge throttling, either intentionally or inadvertently with basin control valves. This will reduce airflow and power, although not as efficiently as other techniques.

Throttling control using an inlet valve is preferred over discharge throttling. Inlet throttling creates a pressure drop and absorbs some of the pressure rise of the blower, moving the operating curve down relative to gauge pressure. The inlet valve pressure drop reduces the air density, further shifting the blower curve and reducing power consumption. Inlet throttling also makes the blower curve steeper.

The most efficient technique for controlling centrifugal blower airflow is variable-speed operation, which is 15% to 20% more efficient than throttling and 5% to 10% more efficient than guide vanes.

The affinity laws predict the performance of centrifugal blowers operating at variable speed and constant inlet conditions.

$$q_{v2} = q_{v1} \cdot \left(\frac{N_2}{N_1}\right)^2 \tag{8.8}$$

$$X_2 = X_2 \cdot \left(\frac{N_2}{N_1}\right)^2 \tag{8.9}$$

$$p_{d2} = p_i \cdot (X_2 + 1)^{3.53} \tag{8.10}$$

$$p_2 = p_1 \cdot \left(\frac{N_2}{N_1}\right)^3 \tag{8.11}$$

Where

q_{v1}, q_{v2} = volumetric airflow at original and new operating speed, m^3/h (ICFM);
X_1, X_2 = adiabatic factor at original and new operating speed, dimensionless;
p_{d2} = discharge pressure at new operating speed, kPa (psia);
p_i = inlet pressure at both original and new operating speed, kPa (psia);
P_1, P_2 = power at original and new operating speed, kW (horsepower); and
N_1, N_2 = original and new operating speed, rpm.

In evaluating performance at a new operating speed, these equations should be applied to several points on the blower curve for the original speed. The new blower curve should be plotted with the original system curve to determine the new operating flow and power (see Figure 8.1). Applying the equations to a single existing operating point will not correctly identify the actual performance at the new speed.

Not all centrifugal blowers can be controlled successfully using variable speed. Blower curves that are essentially flat, characteristic of radial blade impellers, may not provide adequate speed range or control stability. *Rise to surge* is the difference between discharge pressure at the design point and the discharge pressure at the surge point. A rise to surge of 10 kPa (1.5 psi) at nominal speed and maximum inlet temperature is desirable.

When the header pressure increases, or airflow drops below the blower's allowable minimum, surge occurs. This surge can cause contact between the blower's rotating and stationary components. A blow-off valve can vent the pressure to prevent this contact from occurring.

Geared single-stage blowers may be controlled using variable speed. Large blowers of this type are commonly provided with medium voltage motors (>600 Volts), which increases VFD cost. The economics of medium voltage VFDs must be analyzed carefully.

4.3 Multistage Centrifugal Blowers

The multistage centrifugal blower uses several impellers in series (see Figure 8.4). Centrifugal force pushes air entering at the center of the impeller to the outside diameter. The flow discharged from each impeller is directed by the blower case to the inlet of the impeller in the next stage. Each successive stage increases the pressure above the previous stage. The number of stages is selected to provide the necessary discharge pressure. The last

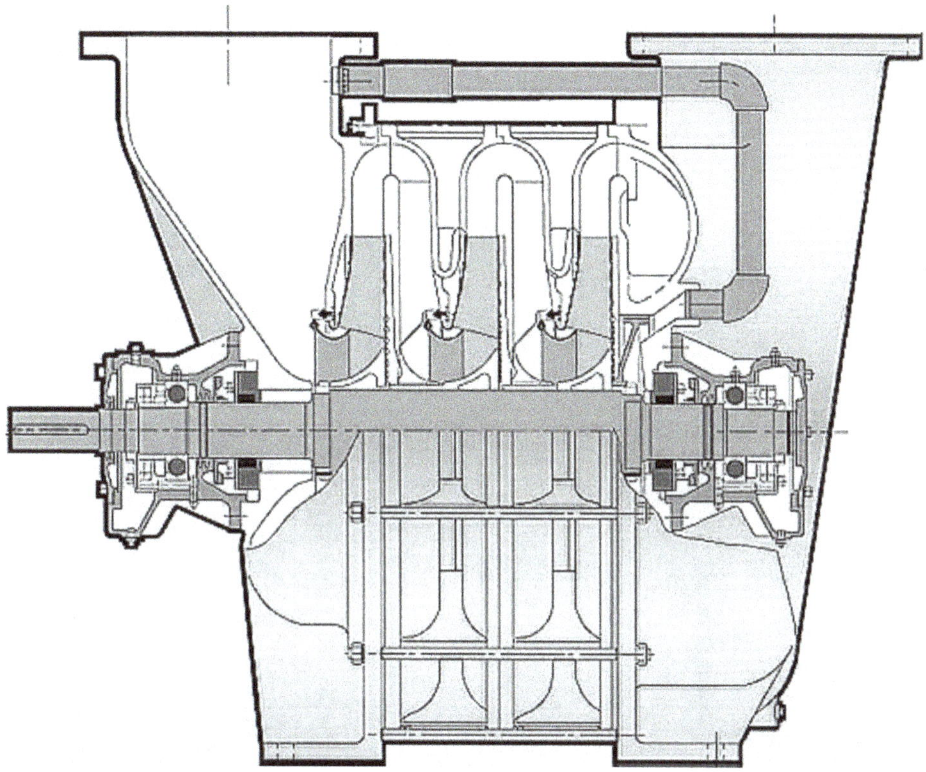

FIGURE 8.4. Multistage centrifugal blower. Reprinted with permission from Continental Blower, LLC.

stage incorporates a volute to direct the air to the blower discharge and into the system piping.

Horizontally split multistage centrifugal blowers are used in high-flow applications. Inlet guide vanes, similar to those used on single-stage blowers, are a common control technique with this type of blower.

4.4 Geared Single-Stage Centrifugal Blowers

Single-stage centrifugal blowers use a single impeller rotating at high speeds to provide the necessary discharge pressure. There are two types of single-stage blowers currently available. The most common design for large blowers (see Figure 8.5) uses gearing between the motor and blower shafts to increase impeller speed above motor speed. Pressure lubrication supplies oil to journal bearings on the blower and/or the gear boxes.

Capacity control and power reduction for geared single-stage blowers are typically provided by adjustable inlet guide vanes (IGVs) or variable discharge diffuser vanes (DDVs) (see Figure 8.6).

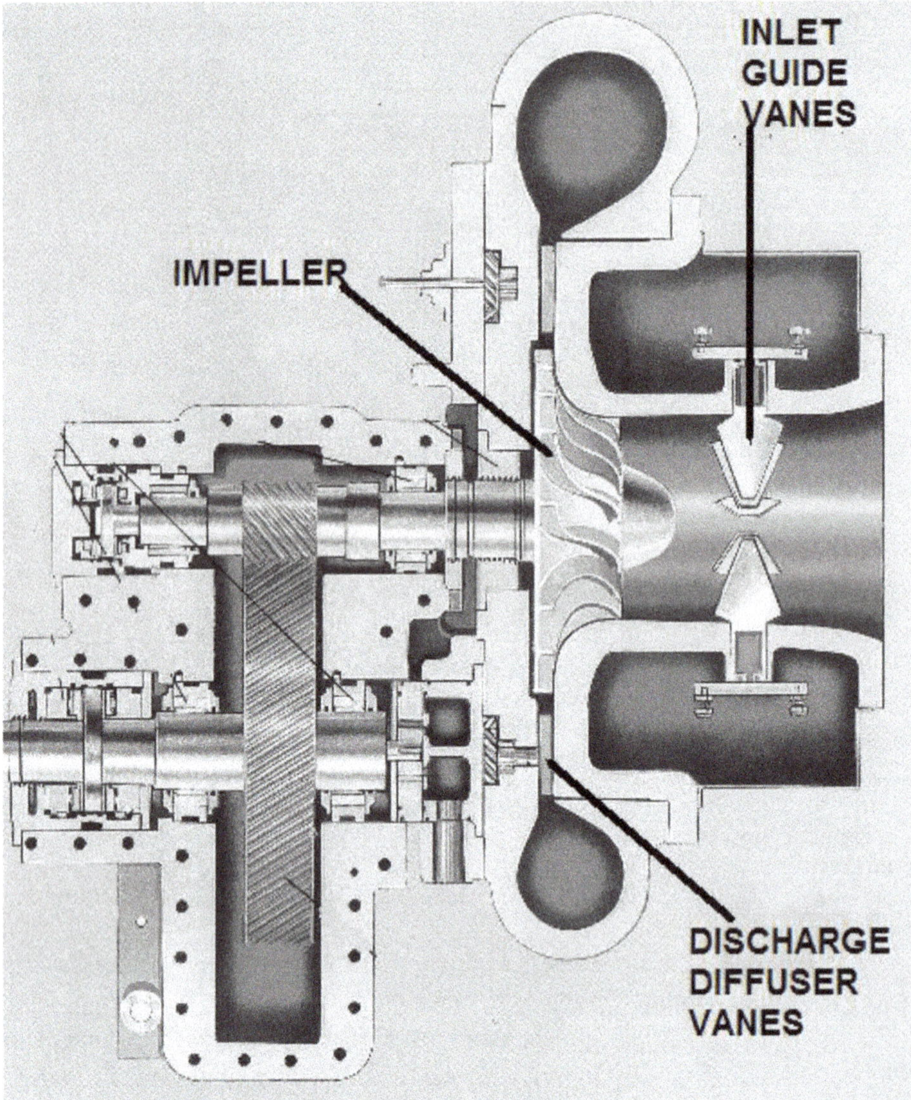

FIGURE 8.5. Geared single-stage centrifugal blower. Reprinted with permission from Howden USA.

An IGV swirls the inlet air in the direction of impeller rotation before the air entering the eye of the impeller. This changes the head of the blower and results in reduced discharge pressure and lower energy requirements at a given airflow rate (see Figure 8.7). Reducing the IGV opening effectively shifts the blower curve down and to the left. At reduced openings, the guide vanes also act as throttling devices.

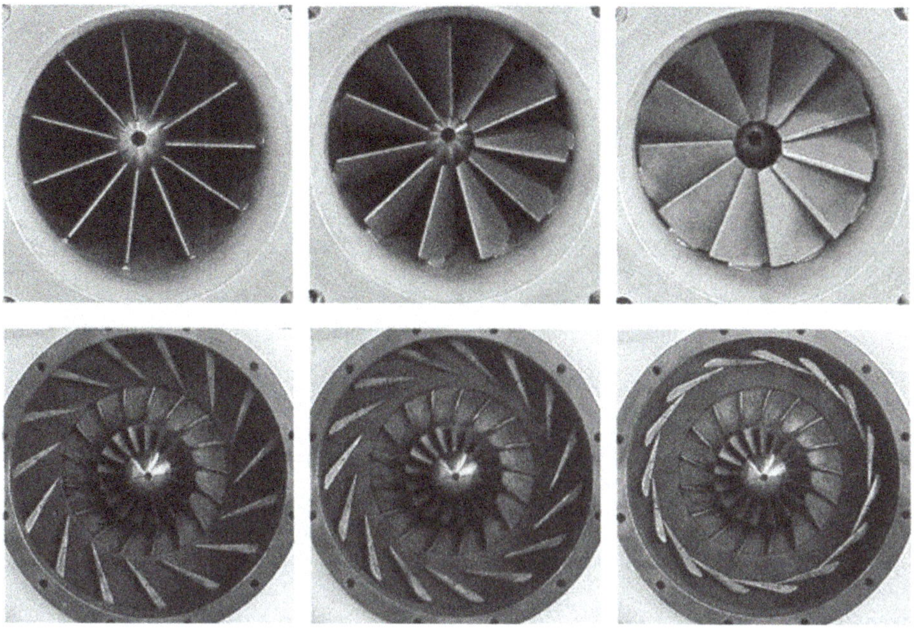

FIGURE 8.6. Examples of inlet guide vanes (top) and discharge diffuser vanes (bottom). Reprinted with permission from Howden USA.

A DDV (see Figure 8.8) changes the conversion of the kinetic energy of the air exiting the impeller to potential energy. As the DDV is closed, the blower curve is shifted to the left, reducing flowrate and power consumption at a given discharge pressure. Some blowers use both IGV and DDV control with proprietary algorithms to coordinate the vanes.

4.5 Direct Drive (Turbo) Single-Stage Centrifugal Blowers

Another single-stage blower design has the impeller mounted directly on the shaft of a special motor operating at high speeds (see Figure 8.9). A special purpose VFD is an integral part of this design. It provides an output frequency much higher than the normal 50 Hz or 60 Hz supplied by utilities. This gearless single-stage blower is often referred to as a *high-speed turbo blower*. Most designs use special bearings to support the blower and motor shaft and do not require external lubrication systems for the bearings.

4.6 Application Considerations

Centrifugal blowers are sensitive to inlet conditions and discharge pressure. Occasionally, an application is encountered in which the blowers will not perform properly during hot weather or periods of high airflow demand. It is

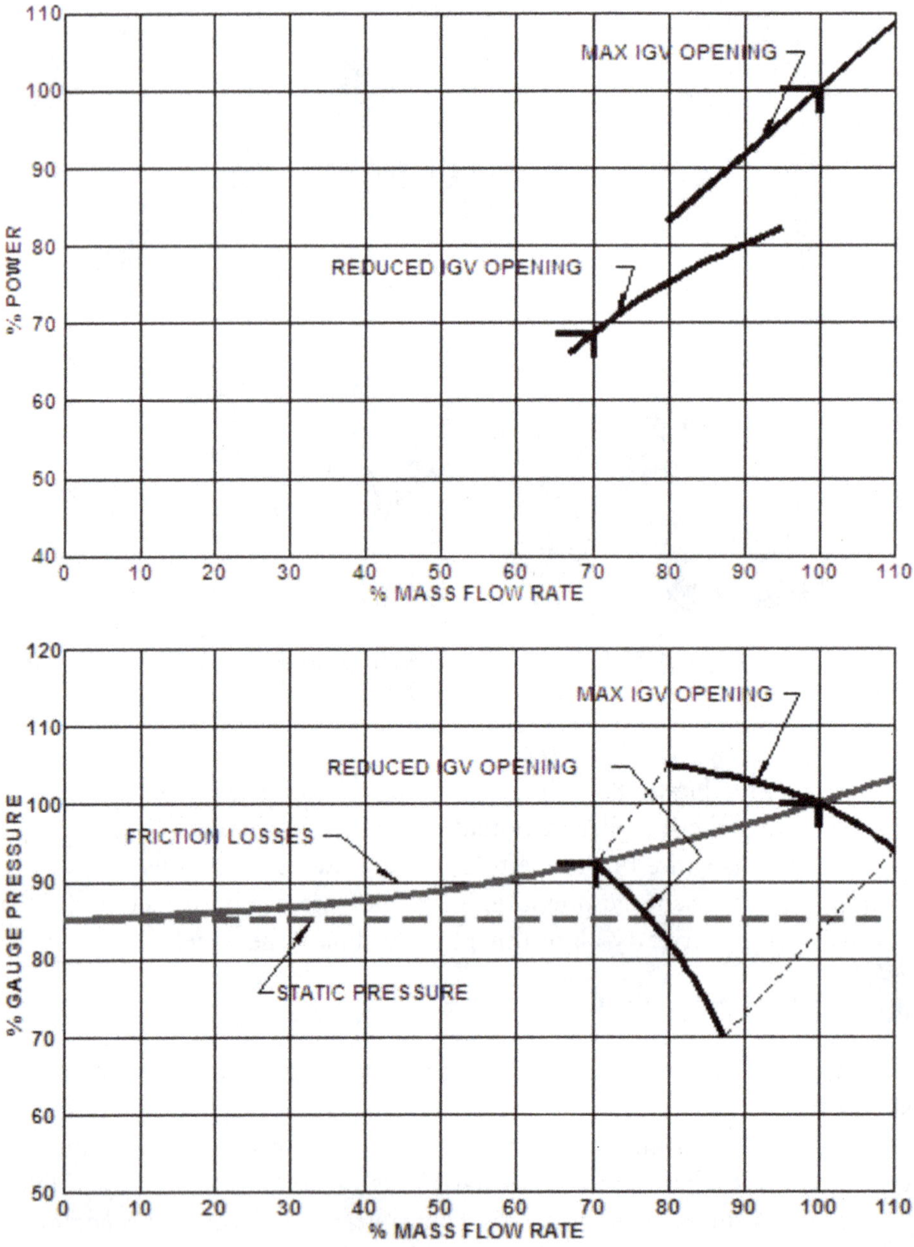

FIGURE 8.7. Inlet guide vane control. Reprinted with permission from Howden USA.

much more common, however, to encounter blowers whose specified design conditions are worse than the requirements of normal operation. Motors for centrifugal blowers are often sized to be "non-overloading" (i.e., the motor nameplate horsepower will not be exceeded at the worst-case design conditions). Efficiency is greatly reduced if the specified discharge pressure is significantly higher than normal operating pressure.

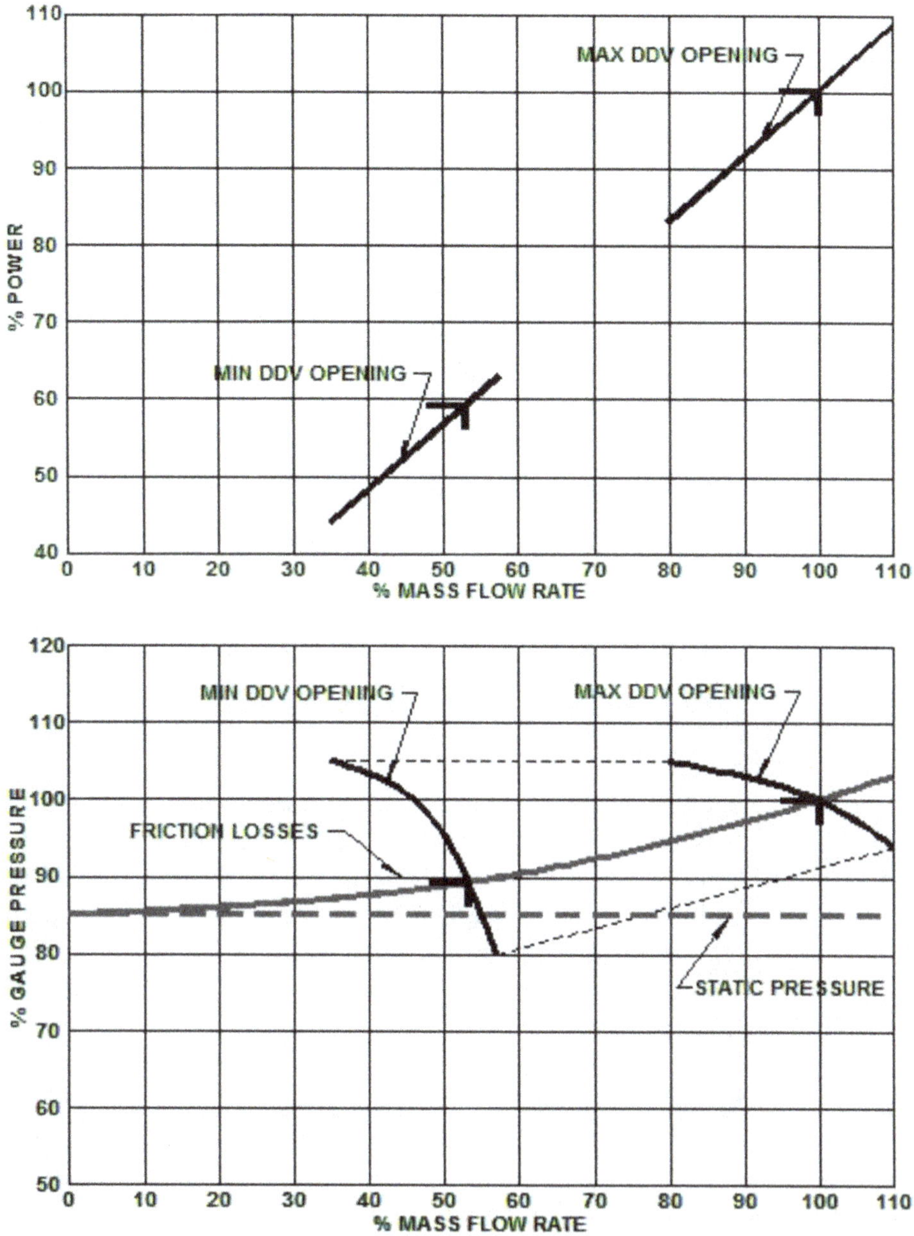

FIGURE 8.8. Discharge diffuser vane control. Reprinted with permission from Howden USA.

4.7 Operating Limits

Centrifugal blowers are limited by the flow and pressure ranges available for a specific blower. The minimum airflow for a centrifugal blower is referred to as the *surge point*. An unstable pulsating flow occurs when operating below this point. The maximum pressure available from the blower typically

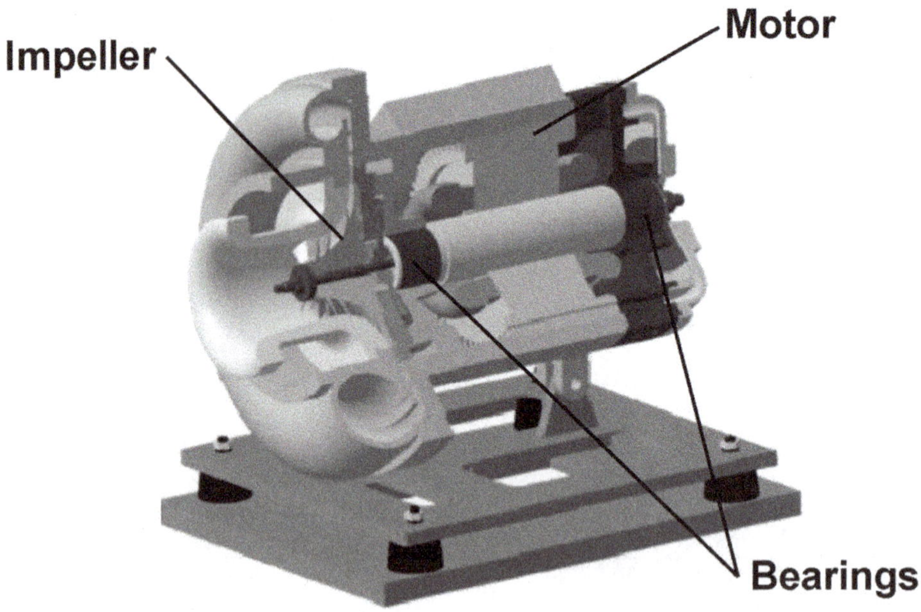

FIGURE 8.9. Direct drive (turbo) single-stage centrifugal blower. Courtesy of Lone Star Blower.

occurs at or near the surge point. The maximum flow may be limited by motor power or system discharge pressure. Centrifugal blower operation may also be limited by the maximum discharge air temperature.

Single-stage blowers are more sensitive to surge than multistage blowers. A blow-off valve is typically provided with single-stage blowers to relieve surge. Like PD blowers, opening of the blow-off valve should be minimized to avoid wasting power.

Single-stage blowers may generate high noise levels because of the high impeller tip velocity. Sound containment enclosures and silencers on inlet, discharge, and blow-off piping are used to minimize the exposure of operators to noise. The power required for enclosure ventilation should be included when evaluating blower power consumption.

5.0 PERFORMANCE TESTING

It is often necessary to test a blower system to verify that its performance matches the specified requirements. Testing is used to verify that pressure, flow, and power characteristics meet expectations. Energy incentive programs often include testing as part of their measurement and verification requirements.

5.1 Field Testing

Testing is sometimes conducted in the field, but this is challenging. Field conditions may not provide the necessary piping configuration for accurate flow measurement or allow operation at conditions comparable to the specified conditions. The accuracy of field instruments may be questionable, and all needed measurements may not be available. In analyzing and reporting field testing results, the limitations and uncertainties of the specific system should be identified and addressed.

At a minimum, field testing should include measurement of ambient conditions, blower system power demand, airflow, and discharge pressure. If accurate airflow measurement is not available, surrogate measurements such as amperage may be used. Airflow calculated from thermodynamic characteristics and power demand at the blower shaft may also be used.

$$\eta_b = \frac{X \cdot T_i}{T_d - T_i} \tag{8.12}$$

$$q_v = \frac{\eta_b \cdot p_b}{K \cdot X \cdot p_i} \tag{8.13}$$

Where

T_d = discharge temperature, K (°R);
T_i = inlet temperature, K (°R);
X = adiabatic factor, dimensionless (see Equation 8.4);
q_v = volumetric airflow, m^3/h (ICFM);
p_i = inlet pressure at both original and new operating speed, kPa (psia);
P_b = blower power, kW (horsepower);
η_b = efficiency of the blower only, decimal; and
K = conversion factor
9.816×10^{-4} (S.I./metric)
0.01542 (U.S. customary).

Because VFDs produce an output waveform that is not truly sinusoidal, typical hand-held instruments for measuring voltage and current may not provide valid measurements, and the VFD power measurement should be used.

5.2 Shop Testing

When possible, performance testing should be conducted using a test stand designed and maintained to provide accurate results. Rigorous calculation procedures can be used for predicting field performance from test data. Instruments should be regularly calibrated. The availability of experienced personnel helps ensure proper procedures and reporting.

Shop testing may involve some differences between the test system and the final configuration. For example, VFDs or motor starters may be contractor supplied and not available. Silencers, filters, or isolation valves may not be installed for the shop test. All differences between test and final systems should be noted in the test report and compensated for in calculations.

5.2.1 Power Guarantees

Many specifications include performance guarantees for blower systems. These range from simple verification of flow and pressure capabilities at a single specified design point to testing at multiple operating points based on anticipated duty cycles and variable process demand. The level of testing required should be related to the size and complexity of the blower system. It is not economically viable to specify elaborate multipoint testing on small blowers or those used for intermittent, short duration operation.

A guarantee specification should include the temperature, discharge pressure, barometric pressure, and relative humidity to be used for the evaluation. For power guarantees, an evaluation at average site conditions is most reflective of actual operating energy. A complete list of the blower system components to be incorporated in the testing should be included.

Manufacturers' standard or advertised performance values include tolerances on performance. The specification should identify the tolerances, if any, that are to be applied in comparing test results to guarantee values. To avoid confusion, it is recommended that zero negative tolerance be applied to specified flow and pressure values, and zero positive tolerance be applied to specified power values. The manufacturer should apply needed tolerances before stating guarantee values.

5.2.2 Performance Projections

It is unlikely that the ambient air conditions encountered during testing will exactly match the specified conditions. The performance at alternate conditions can be predicted using the appropriate thermodynamic calculations identified in the test codes. These calculations can also be used to predict performance from advertised values to site-specific values.

5.2.3 Codes

The complete definition of test procedures, test instrumentation, and calculation methodology for testing blower performance and energy requirements involves considerable detail. There are several performance test codes that provide this detail, and it is recommended that the appropriate code be referenced in specifications or purchasing documents. Among the codes used for defining evaluation procedures are the following:

ASME PTC 13-2018 Wire-to-Air Performance Test Code for Blower Systems
CAGI/PNEUROP BL 300-2016 Low Pressure Air Compressor Packages
ISO 5389:2005 Turbocompressors—Performance Test Code
ISO 1217:2009 Displacement Compressors—Acceptance Tests

6.0 BLOWER SYSTEM RETROFIT OPPORTUNITIES

6.1 Automatic Controls

The variation in aeration system demand makes automatic control of blower airflow cost-effective in most applications. Significant energy reductions are achievable by matching blower airflow to process requirements (see Chapter 9).

6.2 Additional Energy Conservation Measures

Premium efficiency motors should be used with blowers because of their significance to total facility energy consumption. Older blower motors may have low efficiency, and it will generally be cost-effective to replace them with high-efficiency motors.

Both PD and centrifugal blowers can be retrofitted with VFDs to reduce energy consumption. Operating limits should be checked for all blowers. Adequate rise to surge should be verified for all centrifugal blowers.

Substantial improvement in centrifugal blower energy performance can be achieved by changing impellers to more closely match actual operation. In some cases, it is cost-effective to completely replace some blowers with smaller or more efficient units.

7.0 IMPLEMENTATION

Proper implementation is of critical importance for an effective treatment process and for ensuring maximum conservation of energy. Blowers are one element of a complex system and must be matched with control valves, instrumentation, diffusers, and so forth, to consume the least amount of energy. A properly tuned control system will reduce the energy consumed.

7.1 Right Sizing Equipment

A few guidelines provide direction for sizing all types of blowers:

1. Match blower airflow to the process requirements,
2. Minimize system discharge pressure and inlet losses, and
3. Provide flexibility and adequate turndown in the blowers to accommodate loading variations.

7.1.1 Airflow Demand/Duty Cycle

Blower power requirements are directly proportional to the airflow rate. Most treatment processes exhibit wide fluctuations in duty cycle and airflow demand because of several factors. The two most significant are diurnal variations in loading and the difference between operating and design loads. Diurnal load changes occur in most municipal treatment facilities. It is common for facilities to have a ratio of maximum to minimum daily flow of 2:1, and the minimum daily flow is typically 70% of average daily flow.

7.1.2 Blower Selection

Selecting the design airflow rate for blowers while optimizing energy use is a complex task. Minimizing equipment cost and obtaining maximum design-point efficiency suggests using a few large blowers. However, the need to provide reduced airflow for extended periods of time suggests using several small blowers. In general, the energy reduction achievable with several small blowers operating over an extended period will offset the initial savings obtained by using a few large blowers. In either case, if the blowers cannot provide enough turndown, it will be necessary to use excess energy to operate the blowers at their minimum safe airflow rate.

To accommodate airflow demand variations, adequate blower turndown should be provided. Blower turndown is the ratio between maximum and minimum flowrates, as follows:

$$\text{Turndown\%} = \frac{q_{\text{max}} - q_{\text{min}}}{q_{\text{max}}} \qquad (8.14)$$

Where

q = airflow rate, standard or volumetric.

A total system turndown of 80% (5:1) is desirable for most applications to accommodate growth projections and diurnal variations. This turndown is typically obtained by combining changes in the number of blowers operating in parallel and taking advantage of the turndown available from individual blowers.

Blowers can typically provide at least 50% turndown. With 50% turndown, if one blower cannot meet the process airflow needs, a second blower can be started and both blowers operated near minimum flow. This reduces the step change in airflow rate and allows closer matching of the supply to the process demand, thereby minimizing blower power.

Municipal facilities require a standby blower so that 100% of design airflow is available with any one blower out of service. Standby capacity and turndown can be achieved by providing four blowers, each sized to provide 33% of design airflow rate. Providing four blowers, with two sized to provide 50% of design airflow rate and two sized to provide 25% of design airflow rate, is another configuration that provides good turndown. Using a blower configuration with good turndown will directly affect energy consumption.

Another effective strategy is using large blowers to provide the baseline or minimum airflow and smaller blowers to adjust or "trim" the flow. The trim blowers are sometimes referred to as "jockey" blowers. When the large blowers are sized to operate at their BEP, this arrangement may make it possible to maximize the overall efficiency of the process. This approach is common with large systems.

Using blowers of different technologies may also be advantageous, because different technologies may provide higher efficiency at varying operating conditions.

7.2 Operational and Other Considerations

During initial commissioning, facility expansion, or maintenance, there may be short-term abnormal situations. Effects to blowers often include requirements for manual operation, using temporary piping, or other operational issues. Careful planning and coordination between facility management and staff, contractors, consulting engineers, and manufacturers are necessary to minimize the effects of these challenges and interruptions. However, some effect on energy consumption may be expected. Most utilities will waive demand charges during commissioning if they receive notice before the commissioning.

7.2.1 Preventive Maintenance

Preventive maintenance is a cost-effective way to optimize blower performance and energy savings and prevent unnecessary downtime. Even simple items such as changing air filters and oil, greasing motor bearings, and ensuring proper tensioning of belt drives have a significant effect. Consult the equipment operating manual for recommended preventive maintenance.

Inlet filters should be cleaned regularly and monitored for excessive pressure drop. Inlet filter removal efficiency often exceeds equipment requirements and should be checked. Field experience has demonstrated that inlet filters capturing 95% of particles 10 micron (0.000039 in.) or larger are adequate.

7.3 Life-Cycle Analysis

As discussed throughout this chapter, there are many considerations for maximizing the economy of blowers, including intelligent sizing, selection, implementation, and maintenance.

Life-cycle analysis is the process of assessing and comparing the environmental effects of a product or system from "cradle to grave." Although capital cost and maintenance expense can be significant, the most substantial effect of blowers is the energy costs over the lifetime of the equipment. In energy evaluations, it is important to consider realistic operating conditions as well as how the operation may change over time.

8.0 SUGGESTED READINGS

American Society of Mechanical Engineers. (2018). *Wire-to-air performance test code for blower systems* (ASME PTC 13-2018).

Cantwell, J. (2007). Value of energy savings at small facilities. *Proceedings of the 80th Annual Water Environment Federation Technical Exhibition and Conference;* San Diego, California, Oct 13–17; Water Environment Federation.

Compressed Air and Gas Institute. (1973). *Compressed air and gas handbook.*

Dresser Roots. (2004). *Product data book.* Dresser Inc., Dresser Roots: Connersville, Indiana.

Gartmann, H. (Ed). (1970). *De Laval engineering handbook.* McGraw-Hill.

Jenkins, T. E. (2014). *Aeration control system design.* John Wiley & Sons.

Moore, R. L. (1989). *Control of centrifugal compressors.* Instrument Society of America: Research Triangle Park, North Carolina.

U.S. Environmental Protection Agency. (1985). *Summary report, fine pore (fine bubble) aeration systems.* U.S. Environmental Protection Agency, Water Engineering Research Laboratory.

U.S. Environmental Protection Agency. (1989). *Design manual fine pore aeration systems* (EPA/625/1-89/023). U.S. Environmental Protection Agency, Office of Research and Development.

9

Process Controls

Bill McConnell, PE; Alexandra Doody, PE; Alexandra Bowen, PE; Maureen Neville, PE; & Mark Cusac, PE

1.0 INTRODUCTION

Chapters 7 and 8 described the mechanical equipment and systems that may be implemented to reduce energy consumption for aeration. Efficient aeration systems and blower equipment are indeed essential. However, without appropriate process controls, the energy efficiency of these components cannot be optimized and may even be compromised. This chapter focuses primarily on process controls that can be used to optimize process performance while reducing energy usage for major energy consumers in a typical water resource recovery facility (WRRF).

The type of process control components required are described first, including a description of analyzers necessary to implement process control. Parameter-based aeration control approaches, from simple (and antiquated) dissolved oxygen monitoring with manual adjustment, to more advanced systems such as ammonia-based aeration control (ABAC) are described. Aeration control systems that use modulating valves are also discussed, including various most open valve (MOV) control schemes.

Though this chapter focuses primarily on control approaches related to aeration, ultraviolet disinfection is another energy-intensive unit process, and process control approaches to reduce energy consumption of ultraviolet systems are also described in this chapter. While focusing primarily on aeration and ultraviolet disinfection, it is acknowledged that this chapter is not comprehensive regarding opportunities to reduce energy consumption or improve energy efficiency by implementing process control systems at WRRFs. Significant energy-intensive equipment, such as pumps and mixers, is typically and rightly controlled to meet process objectives and these process-based goals outweigh control for managing energy use. Control of pumping systems for return activated sludge, internal mixed liquor recycle, and so forth, based on process requirements is covered extensively in other manuals. Mixer settings are generally established during commissioning and then not modified unless process goals—such as on/off operation to improve enhanced biological phosphorus removal—dictate. Other unit processes, such as aerobic and anaerobic digestion, can be optimized for energy usage or production by the implementation of appropriate process controls; however, these are addressed elsewhere and are beyond the scope of this chapter and this Manual of Practice.

2.0 PROCESS CONTROL COMPONENTS

2.1 Introduction

This section describes the components used at WRRFs as tools for implementing efficient process operation. Process instrumentation and control valves are described given their vital importance in process control.

2.2 Process Analyzers

Process analyzers are used to provide online measurement of water quality parameters used for process control and control of aeration systems. See Water Environment Federation's *Automation of Water Resource Recovery Facilities Manual of Practice No. 21* (WEF, 2013a) for a detailed discussion on analyzers, their theory of operation, and their uses.

2.2.1 Analyzer Types

2.2.1.1 Dissolved Oxygen

Dissolved oxygen measurement is the primary control parameter for aeration control. Dissolved oxygen is the amount of free oxygen dissolved in water. It is typically expressed in milligrams per liter. A local dissolved oxygen setpoint is typically the basis of a feedback control for aeration processes, even when that dissolved oxygen setpoint is dynamic and calculated based on other water quality parameters. See ABAC schemes, which are discussed later in this chapter. Two main types of dissolved oxygen probes are typically used at WRRFs:

- **Ion selective electrodes** (ISEs)—These probes use a semipermeable membrane that allows oxygen to pass through it but not water or other electrolytes. An electrochemical cell is set up on the other side of the membrane, and as oxygen is reduced by the cell, it produces a measurement correlated to oxygen concentration in the water.

Oxygen-permeable ISE probes have been traditionally used in aeration process control, but luminescent probes have seen increasing use more recently because of advances in reliability and reduced maintenance requirements.

- **Luminescent**—Luminescent probes use an oxygen-permeable polymer layer that is infused with a material that luminesces red light when hit with either blue or green light from an LED source. A sensor then detects the amount of red light. As the dissolved oxygen concentration increases, less red light reaches the permeable layer because oxygen in solution quenches the reaction. Luminescent probes are typically slightly less accurate than membrane probes in ideal circumstances, but they often require less cleaning and show less drift because of fouling.

Dissolved oxygen sensors are typically placed at strategic points in an aeration basin (e.g., one-quarter, halfway, and/or three-quarters through the process depending on the control configuration) to adequately control the process. Depending on the setup of the treatment facility, it may make sense to install dissolved oxygen sensors before the aeration basin to monitor anoxic or anaerobic conditions and help combat dissolved oxygen sag. A sensor installed toward the effluent end of the process can be used to determine whether the process is over-aerated and, therefore, wasting energy.

2.2.1.2 Oxidation–Reduction Potential

Oxidation–reduction potential (ORP) analyzers measure the water's potential to oxidize or reduce compounds in water. The potential is generated by the oxidant (such as chlorine) or reductant (such as sulfite) in the wastewater stream.

Oxidation–reduction potential is a potentiometric measurement like pH. The potential (tendency) of a medium to transfer electrons is sensed by an inert metal electrode and read relative to a reference electrode immersed in the same medium.

Unlike pH, dissolved oxygen, or other chemical measurements described here, ORP is not a concentration measurement. Instead, it notes the effect of the concentration of all dissolved species in the medium. For this reason, it cannot be used to directly measure the effect of an increase in the concentration of oxygen or other individual species in the wastewater. As an oxidant, the addition of oxygen will tend to increase the ORP reading. Unless a pre-oxidation ORP measurement is made in the same water, however, this is unlikely to be of use. Typically, the *change* in ORP reading is more useful than any individual reading at a point in time.

Oxidation–reduction potential can be used for biological nutrient removal (BNR) processes in which operators desire to monitor anaerobic versus anoxic conditions. Oxidation–reduction potential values in the range of −100 to +100 mV indicated anoxic conditions suitable for denitrification, while ORP values less than −150 mV indicate anaerobic conditions. Theoretically, ORP measurements can help operators to make process decisions regarding mixed liquor recycle (MLR) pumping rates and aeration intensity, which can have an energy effect. However, in the authors' experience, ORP measurements in mixed liquor can be difficult to maintain in calibration. Maintenance considerations associated with using ORP probes in mixed liquor is likely one reason why ORP is not more widely used for automated control of BNR processes.

2.2.1.3 Nitrogen

2.2.1.3.1 Ammonia

Ammonia is typically monitored in the following places:

- Facility influent, to allow load balancing and feed-forward control of aeration;
- Aeration tanks, to monitor nitrification and optimize aeration control;
- Facility effluent, to monitor compliance with permit requirements;
- Intermittent aeration systems, to control aeration time based on mixed liquor ammonia levels;
- Supernatant of dewatered biosolids, to monitor the high ammonia levels returning to the head of the facility; and
- Sidestream treatment processes that remove nitrogen.

Two different types of instruments are available for real-time monitoring of ammonium within a municipal WRRF, as summarized in the brief descriptions below:

- **Ion selective electrodes** (ISE)—ISE probes are immersed in a process tank and measure a millivolt (mV) potential across a membrane that selectively adheres NH_4^+ ion. A site-specific calibration curve is used to correlate the mV readings to an NH_4^+ concentration.
- **Wet chemistry analyzers**—Wet chemistry colorimetric analyzers require a sample pump to move the water sample from the process

tank, through a filter, and then into the analyzer. Reagents are used to react with ammonium in solution, and the resulting color is correlated to an NH_4^+ concentration.

Ammonia instruments can be used either for monitoring purposes only, or for automated process and aeration control. Aeration control strategies that use ammonium as the primary control variable are referred to as ABAC schemes, which are discussed later in this chapter.

2.2.1.3.2 Nitrate

Nitrate (NO_3) is a stable form of oxidized form of nitrogen that can be present in wastewater after biological treatment. Nitrate is typically monitored in the following locations:

- The collection system, to control the feed of odor control chemicals such as calcium nitrate;
- The first anoxic zone, to control the mixed liquor recycle rate;
- The aeration basin, to control intermittent aeration systems;
- The second anoxic zone or denitrification filters, to control supplemental carbon feed;
- The facility effluent, for compliance assurance; and
- The sidestream treatment systems that remove nitrogen.

Two different types of NO_3 probes as summarized below:

- **Ultraviolet Absorbance**—These probes are more typically used to monitor nitrate. In water, nitrate absorbs ultraviolet light between 215 and 240 nm. Online ultraviolet absorbance readings are converted to a nitrate value, typically displayed as milligrams per liter of NOx-N, because both nitrate and nitrite absorb light in that spectrum. Advantages of the ultraviolet absorbance method are the absence of reagents and the ability to directly measure nitrate in the process.
- **Ion-Selective Electrode**—Another nitrate monitoring method involves ISE technology, similar to what was described for dissolved oxygen or ammonia sensor, but using a membrane treated with a compound so that it is selective to nitrate.

Ultraviolet absorbance or ISE probes can be mounted directly in the process.

2.2.1.3.3 Suspended Solids

Suspended solids probes used in aeration basins can be used to monitor mixed liquor suspended solids (MLSS) concentrations in the bioreactors, thus allowing operators to evaluate biomass inventory under aeration in real time. Additionally, some proprietary ABAC controllers using "feed-forward" strategies rely on real-time MLSS information in their control schemes. Suspended-solids probes can also be used to monitor the concentration of return activated sludge, and enable automatic control of biomass inventory and solids retention time.

2.2.1.4 Offgas Oxygen Content

Another instrument that can be used to monitor, and in some cases control, aeration for activated sludge systems is a hood placed on the surface of the mixed liquor to measure the oxygen transfer rate. Offgas, collected under the hood, is compared to atmospheric air to quantify the amount of oxygen absorbed in the liquid. Oxygen is measured with a fuel cell, and carbon dioxide and relative humidity are also measured relative to ambient air. The offgas flowrate is also measured. These measurements allow for estimation of real-time field oxygen transfer efficiency (OTE), and when combined with information about the airflow and dissolved oxygen concentrations, the oxygen uptake rate or oxygen transfer rate can also be calculated. The degree of fouling that affects the alpha coefficient (alphaF) can also be estimated if clean water OTE testing was conducted on the diffusers before commissioning. Thus, offgas hoods used for monitoring purposes can allow operators to evaluate when diffuser cleaning is merited. Some WRRFs in Europe also use offgas hoods for real-time aeration control, where the real-time OTE information is used to calculate the airflow required in real time; however, this approach is not commonly used and thus is not discussed further in Section 4.0. One potential advantage of using offgas hoods for process control is that they can reportedly respond to changes in load more quickly, thus resulting in tighter dissolved oxygen control and more stable process performance. Figure 9.1 presents a schematic of an offgas analysis system using a hood and Figure 9.2 is a photograph of an offgas hood in service.

2.2.2 Implementation, Placement, and Operations and Maintenance Considerations

Identifying proper application and placement is essential for successful use of instruments for real-time process control. Instruments that function well when monitoring secondary clarifier effluent might not perform adequately

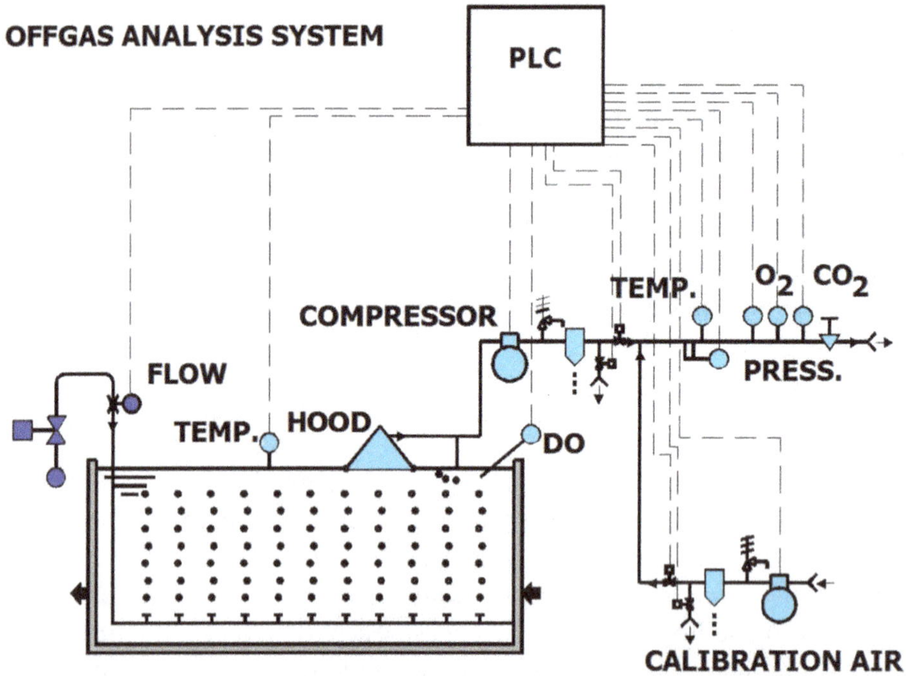

FIGURE 9.1. Schematic of offgas hood design. Reprinted with permission from Advanced Aeration Controls LLC.

FIGURE 9.2. Offgas hood at the Grafton WRRF in Grafton, Wisconsin. Reprinted with permission from Advanced Aeration Controls LLC.

in a mixed liquor or primary effluent environment. Some instruments require more frequent maintenance and calibration than others, and operations personnel should be made aware of these requirements before a significant capital investment is made. Advanced control strategies using instruments that cannot be maintained are doomed to fail. Piloting new instruments for a particular location and purpose at a given WRRF is a good approach to determine fit and improve the likelihood of a successful outcome.

If piloting is not possible, experience from other WRRFs should be considered. Table 9.1 compares the ISE ammonium instrument against the wet chemistry ammonium instrument type with respect to their specifications

TABLE 9.1. Comparison of implementation and placement considerations for different ammonium instrument types (Doody & Neville, 2017).

Type	ISE Probes	Wet Chemistry Analyzers
Range	Nominally capable of 0–1000 mg/L N	Nominally capable of 0.02–1000 mg/L N
	Typically calibrated at approximately 1–20 mg/L N	Typically calibrated at approximately 0.05–20 mg/L N
Accuracy	± 5% of mV signal + 0.2 mg/L	± 3% + 0.05 mg/L
Operator Feedback and Lessons Learned	Locations most frequently used in ABAC facilities are anoxic/anaerobic zone before aerobic zone, or front half of aerobic zone	Location most frequently used in ABAC facilities is end of aerobic zone, secondary effluent, final effluent
	Mixed success in primary effluent location (grease buildup noted as problem for some)	Mixed success in anaerobic/anoxic mixed liquor location (tubing tends to turn black and require more frequent replacement than in downstream locations)
	Poor results when placed at end of aerobic zone unless target effluent NH_4-N > 1 mg/L (probe tends to struggle at low concentrations)	Most suitable for control systems requiring accurate low-level measurement (below 1 mg/L NH_4-N)
	Require frequent accuracy checks and re-calibration to maintain accuracy for control purposes	Accuracy checks are recommended to determine when cleaning of internal tubing or flow cells is required (readings will tend to run low if these items are dirty)
	Replacement cartridge heads are expensive, and some are less reliable than others out of the box	Reagent cost can be reduced by increasing the time interval between measurements, but this must be balanced with needs of process control system

and based on WRRF operator feedback on their function and accuracy in conditions at municipal WRRFs (Doody & Neville, 2017).

2.3 Airflow Control Equipment

Some facilities, particularly larger WRRFs and WRRFs requiring precise process and dissolved oxygen control to optimize BNR performance for stringent effluent requirements, opt to implement modulating valves to control the distribution of airflow to different tanks or even to every diffuser drop leg. Using modulating valves allows the aeration system to provide the right amount of air to different locations, which often differ because of imbalances in mixed liquor concentration, uneven influent flow splits, and differences in pipe installation or diffuser fouling rates among different tanks.

Proper selection of appropriate airflow control equipment is an integral component of the performance and efficiency of an aeration system. The primary components of airflow control equipment include airflow control valves and airflow meters. Properly sized flow control valves installed along aeration piping coupled with strategically placed airflow meters are critical to achieving and maintaining an aeration system's selected parameter setpoint(s) throughout the aeration tanks.

2.3.1 Flow Metering

Historically, airflow metering was only used on the main header to measure the total airflow from each process blower for the sole purpose of monitoring blower performance. This approach did not provide any level of control to the airflow to each basin. Because of increased efforts to minimize energy consumption and improving overall aeration system efficiencies, more extensive airflow metering methods have been used to either balance airflow delivery between basins or to control airflow delivery to specific zones along the length of aeration tanks. Airflow meter signals (or setpoints) can be integrated into control loops at WRRFs to automate the aeration system in response to variable influent loading characteristics.

Flow meters should always be located upstream of flow control valves to minimize disturbance to the flow signal. Each type of flow metering technology has recommended installation requirements to ensure they achieve optimal performance, accuracy, and repeatability. Wide variations in actual field conditions and installation constraints (notably inadequate lengths of straight runs of aeration piping for establishing a uniform velocity profile across the pipe diameter) may differ from ideal conditions under which flow meters were calibrated. It is critical to maintain appropriate straight lengths of pipe upstream and downstream of the flow meters.

Thermal mass flow meters measure mass flow directly, without the need for pressure or temperature compensation, and induce low pressure loss across the flow element, itself. Insertion thermal mass flow meters can be installed directly into aeration piping via a single tap, making them amenable to retrofit applications. Thermal mass flow meters can achieve a high turndown ratio (100:1), allowing them to measure low flows precisely and respond quickly to fluctuating airflows. Older analog thermal mass flow meters had lower accuracy, but newer digital signal processing technology has resulted in improvements to long-term offset stability, accuracy, and resolution (Lipták & Venczel, 2016). Some manufacturers provide multipoint thermal mass sensors, which can provide better accuracy than a single thermal mass flow meter, particularly when shorter pipe lengths are available. Considering all factors, including accuracy, permanent pressure loss, and cost, thermal mass flow meters should be considered the strongest choice for aeration flow measurement.

If a piping layout is inadequate, there are flow conditioning devices and flow straighteners that can be used to mitigate the straight pipe requirement, to some extent. These designs include averaging pitot tubes, flow tubes, and flow straighteners.

2.3.2 Flow Control Valves

Within the aeration system, flow control valves are used to balance the airflow into the aeration zones. Flow control valves are used to create a pressure differential to control airflow rates and are installed downstream of flow meters. The airflow split from the blower to each aeration tank, zone, and diffuser grid and be controlled by adjusting the position, or throttling, of flow control valves. Air flow control valve positions can be adjusted manually or can be automated based on various control parameters. When automated, a control valve is equipped with an actuator that acts within a control loop.

Different airflow control valve types can be used to accomplish varying degrees of precision, which can be used to fine-tune and maintain suitable dissolved oxygen concentrations down the length of bioreactors, thereby improving overall process control for a biological process. When airflow is restricted, and a valve position closes, a greater pressure differential across the valve is induced.

The flow coefficient, C_v, is defined as the flow (in gpm) through a valve that induces a pressure differential of 1.0 psi. This flow coefficient increases as valve size (diameter) increases and as a valve moves open. Flow can be calculated using C_v as follows:

$$Q_s = 22.66 \cdot C_v \cdot \sqrt{\frac{p_u \cdot \Delta p_{vi}}{SG - T_u}}$$

Where

Q_s = airflow rate, SCFM;
C_v = valve flow coefficient from manufacturer's data, dimensionless;
p_u = upstream absolute air pressure, psia;
Δp_v = pressure drop (differential) across the valve, psi;
SG = specific gravity, dimensionless (1.0 for air); and
T_u = upstream absolute air temperature, °R.

The International Standard equivalent of C_v is called the flow factor, K_v, which is defined as the flow (in m^3/h), which induces a pressure differential of 100 kPa (1 bar). K_v is proportional to C_v as follows:

$$K_v = 0.862 \cdot C_v \tag{9.1}$$

Furthermore, flow can be calculated as a function of K_v as follows:

$$Q_s = 4.828 \cdot K_v \cdot \sqrt{\frac{p_u \cdot \Delta p_{vi}}{SG - T_u}} \tag{9.2}$$

Where

Q_s = airflow rate, m^3/h;
K_v = valve flow coefficient from manufacturer's data, dimensionless;
p_u = *kPa*
Δp_v = pressure drop (differential) across the valve, *kPa*;
SG = specific gravity, dimensionless (1.0 for air); and
T_u = upstream absolute air temperature, °K.

As shown in these equations, the relationship between C_v (or K_v) and the pressure differential across the valve is non-linear. Therefore, the variation with C_v (or K_v) with valve position is also non-linear for most of valves, which can cause difficulty if the design or airflow controls are not appropriate. For most valve types, there is a loss of control at the upper and lower range of a valve's position range. For this reason, is it important to consider a valve's optimal operating range (position, or percent open) as indicated as the linear portion of each valve's unique characteristic curve. By using flow control valves with lower headloss, aeration blowers will draw less power, which shall result in aeration system power savings.

Inherent characteristic curves are published by valve manufacturers based on tests performed in a system where the pressure drop across the test valve is held constant at all valve openings and flowrates. The results represent the relationship between valve flow capacity and valve opening when there are no system effects involved.

Each flow control valve type requires different installation requirements in relation to flow meter placement, induces different amounts of headloss across the valve, and can achieve varying degrees of precision based on each valve's unique characteristic curve. The degree of precision that each flow control valve type can achieve is ultimately dictated by the level of accuracy afforded by the actuator used to adjust the valve position.

2.3.2.1 *Butterfly Valves*

Butterfly valves (BFVs) are commonly used for control valves in North America. The valve accomplishes control by rotating a disk through 90 degrees. The S-shaped characteristic curve of BFVs results in an error range of valve stokes that provide stable control (generally approximately 50% to 75%). Properly sized BFVs are frequently one size smaller than aeration piping line size, requiring piping layouts to be able to accommodate added lay lengths associated with eccentric reducers. When line-sized BFVs are used at facilities, a small change in valve position results in a large change in flow, causing the valve to constantly "seek" its proper position to maintain a setpoint.

High performance BFVs are designed using a method of pinning the disk to the shaft that minimizes lost motion between the shaft and the disk, as well as a more streamlined disk. These characteristics coupled with precise actuators result in more accurate flow control (reported accuracy to repeated signal changes as small as 0.5% of total signal) and lower headloss (estimated at 1.4 kPa [0.20 psi] under typical conditions) compared to an AWWA butterfly valve. However, the valve characteristic curve has a similar shape as to a standard BFV (narrow range of stable operating control).

Butterfly valves have the lowest capital cost when compared to other control valves. Butterfly valves typically require 5 straight pipe diameters between the flow meter and control valve, because of the air disturbance produced by the central disk.

2.3.2.2 *Knife Gate Valves*

Knife gate valves have a sliding gate that moves up, out of the airflow path, and down into the airflow path. Knife gate valves have a seat along the flow axis that allows the valve to close completely. Knife gate valves have no obstruction in the flow path when in the completely open position, with results in minimal pressure loss across the valve. Knife gate valves provide a more precise regulation of flow than BFVs. More specialized knife gate valves provide a greater operating range with different shaped gates (notably v-port valves). Despite these advantages, knife gate valves are typically more expensive compared to low-cost BFVs.

2.3.2.3 Specialty Valves

There are various specialty air control valves commercially available that are designed to produce a more linear characteristic curve compared to the traditionally used butterfly and knife gate valves. These valves have a more predictable and wider range of valve strokes that can be used to regulate flow because of the more linear characteristic curve. One such specialty valve is formed by six segments that are arranged to form a centralized circular opening, similar to a camera aperture. The six segments slide on each other along convex/concave edges that allow for stepless adjustments in the size of the aperture. This valve has a more predictable and wider range of valve strokes that can be used to regulate flow (a range of approximately 15% to 80% stroke is likely to result in the most stable operation). The centralized circular opening in the valve produces a laminar flow path down the center of the pipe that allows the valve to be installed just downstream of the flow meter (eliminating longer lengths of aeration piping required with BFVs).

The most sophisticated type of control valve currently available is relatively new to the market and engineered specifically for aeration applications requiring precise control. The shape of the valve is unique in that a central control element, shaped like a bullet, is located at the center of a venturi-shaped body, so that flow is directed through a 360-degree free-flow annular orifice. The valve stroke occurs parallel to flow—the control element moves against the incoming airflow to stroke open and it moves with the incoming airflow to stroke closed. The shape of both the venturi body and the control element have been fine-tuned to achieve a near perfect linear relationship between stroke and airflow, which means that the valve can operate stably from 0% to 95% stroke. Because of the venturi shape of the valve and the way it forces air toward the walls of the valve body, pressure builds on the upstream side of the valve. This means that the valve works as a flow conditioner upstream of the control element, resulting in repeatable flow measurement even when the flow meter is positioned only 0.5 pipe diameter upstream of the valve flange.

Specialty valves typically have higher upfront capital costs compared to BFVs and knife gate valves. However, these higher capital costs may be warranted based on increased level of precision achieved by these control valves and potentially decreased system headloss induced across each valve (for particular specialty valve types) as dictated by design objectives and/or life-cycle costs. More precise airflow control is particularly beneficial when trying to meet stringent nutrient limits, especially low nitrogen and phosphorus limits, within the same BNR process. The decreased energy costs offered by lower headloss through certain specialty valves can make those valve types most beneficial over a life-cycle analysis, especially if the power costs at a WRRF are high.

3.0 PROCESS CONTROL CONCEPTS

3.1 Introduction

Optimizing the energy efficiency of the biological treatment process first requires consideration of the process requirements and controls necessary for the WRRF to accommodate the range of influent flows and loads while also consistently meeting effluent discharge requirements.

3.2 Process Considerations

3.2.1 Process Influent Load Variation

Wastewater flows and constituent mass loadings fluctuate diurnally with low flows and loads typically occurring in the early hours of the day, and maximum loads occurring around noon time. From an energy efficiency perspective, it is important for facilities to monitor and understand this variation in flows and loads and adjust operational strategies to treat various loading conditions. If a facility does not respond to low influent flow and load conditions, there is the potential to waste energy.

Variable influent loading results in changes to oxygen demand to which an aeration system must respond. Automated control of the aeration process allows for adjustments to system output so the air delivered to the biological process more accurately matches oxygen demand. Aeration control can also account for changes in OTE that result from ambient conditions or age and condition of equipment, improving the efficiency of the process. Aeration control can result in significant energy savings for a WRRF. The amount of energy savings is site specific and highly dependent on the type and level of control implemented.

3.2.2 Effluent Requirements

The primary goal of a WRRF is, of course, to achieve continuous, reliable compliance with the facility's discharge permit. Measures taken to optimize the process for energy efficiency while achieving that goal should consider whether such measures increase the risk of permit violation. For example, operating in intermittent aeration mode (as described in Section 3.3.2 below) may make compliance with a stringent ammonia-nitrogen limit challenging.

More generally, reducing the operating dissolved oxygen concentration of the process too low may tend toward proliferation of filamentous organisms, perhaps compromising compliance with an effluent total suspended solids limit.

3.3 Process Concepts for Energy Use Reduction

Energy savings can be realized by consideration and selection of a process control strategy that both meets process performance requirements and targets reduction in energy required. Three such process control concepts are described below.

3.3.1 Simultaneous Nitrification/Denitrification

Simultaneous nitrification/denitrification (SND) can be implemented at WRRFs that achieve nitrification, whether or not denitrification is also required to achieve removal of total nitrogen. Generally, the nitrification and denitrification processes occur in separate oxic and anoxic treatment zones. However, at low dissolved oxygen concentrations, it has been demonstrated that both processes can be achieved in the same reactor, which is the SND process. Simultaneous nitrification/denitrification is described in more detail in Chapter 7 of this manual and elsewhere (WEF, 2013b).

Implementing SND has the potential to reduce energy consumption by three primary means: (1) taking advantage of the "oxygen credit" that results from denitrification; (2) operation at lower dissolved oxygen concentrations than traditional oxic processes; and (3) elimination of internal MLR pumping systems that would otherwise demand power.

- Denitrification occurs when nitrate in the mixed liquor, resulting from the nitrification process, is used as the oxygen source to support respiration. In the absence of dissolved oxygen (under anoxic conditions) the heterotrophic microorganisms use the oxygen in the nitrate ion. For every kilogram of nitrate denitrified, 2.86 kg of oxygen is made available to the heterotrophs. This is oxygen that need not be transferred to the mixed liquor by way of aerators or diffused aeration systems, reducing energy requirements.
- As described in Chapter 7, one of the factors that significantly affects the OTE of aerators and air-diffusion systems is the residual dissolved oxygen concentration maintained in the bioreactor. The increase in OTE results in reduced aeration rate or mechanical energy needed to provide the required oxygen. Traditionally, a residual dissolved oxygen concentration of 2.0 mg/L has been the design criterion for aeration systems; however, operating these systems at lower dissolved oxygen—while still providing sufficient oxygen to the process—will result in energy savings.

- Mixed liquor recycle pumping systems are a common component of biological processes designed to achieve denitrification. A common process configuration known as the Modified Ludzack-Ettinger (MLE) process uses an MLR system to pump a high volume of nitrified mixed liquor (typically up to 400% of the influent flowrate) from the effluent end of the oxic bioreactor to a pre-anoxic zone. Though this pumping service is low head, given the high flowrate, the pumping process can use a significant amount of energy. For example, the MLR pumping system for a 37.85 ML/d (10 mgd) WRRF usng 400% MLR pumping system would likely consume approximately 20 to 25 KW. Simultaneous nitrification/denitrification systems do not include such pumping.

3.3.2 Intermittent Aeration

Intermittent aeration, also called cyclic aeration, is similar to SND in that both nitrification and dentification occur in the same process volume; however, rather than operating in a continuous, low dissolved oxygen environment, alternate oxic and anoxic environments are provided in a controlled, timed sequence. Nitrifying activated sludge process can be converted from continuous aeration to intermittent aeration by alternating diffused aeration zones by automatically actuated valves. Intermittent aeration is described in detail elsewhere (WEF, 2013b). Intermittent aeration offers the same potential energy-efficiency benefits as described earlier for SND (i.e., denitrification oxygen credit, reduced aeration requirements (because of the aeration "off" period), and lack of MLR systems.

Intermittent aeration offers a unique opportunity to use the process analyzers described in Section 2.0 to achieve demand-oriented process control. Instead of using a simple timer control, alternating from an oxic period to anoxic conditions, and vice versa, can be automatically controlled based on dissolved oxygen, ORP, and/or the combination of ammonia and nitrate. Dissolved oxygen-based automatic control in this case would simply alternate air ON/OFF based on dissolved oxygen setpoints—with the resulting dissolved oxygen cycling within a preset range. Oxidation–reduction potential-based control would be similar, with air turning ON when the ORP reached a low setpoint (e.g., −100 mV) and turning OFF at high ORP (e.g., +100 mV).

Oxic and anoxic periods can also be controlled based on feedback from on-line ammonia and nitrate analyzers. Conceptually, air would be turned ON when the effluent ammonia concentration increases beyond a high setpoint; and would turn OFF when either the ammonia concentration decreases below a low setpoint and/or the nitrate concentration increases above a high nitrate setpoint. This approach would result in both ammonia and nitrate varying to within a target band.

3.3.3 Mixed Liquor Recycle Pumping Optimization

Many processes that achieve nutrient removal use at least one MLR system. Mixed liquor recycle is typically from the discharge end of an aerobic zone, characterized by nitrified, and therefore nitrate-rich mixed liquor, to an anoxic zone where the nitrate is denitrified. Mixed liquor recycle flow rates can be high—commonly 400% of influent flowrate—so the required pumping systems can consume a significant amount of power.

The process analyzers described in Section 2.0 can be used to optimize MLR rates, and thereby minimize energy use. By continuously measuring nitrate in the anoxic zone, the MLR pumping rate can be turned down during periods of high loading to avoid nitrate breakthrough of the anoxic zone. Such breakthrough would characterize an overloaded anoxic zone, indicating that reduction in pumping rate would not reduce process performance, while allowing for reduced energy use.

4.0 AERATION CONTROL CONCEPTS

4.1 Introduction

Selecting the appropriate type of aeration control for a given activated sludge process requires consideration of the importance of energy or other operational cost savings as well as the process objectives and the effluent quality requirements that the WRRF must meet. Manually adjusted aeration systems require the least cost to implement, but invariably waste energy because the system must be operated at higher aeration rates than are strictly necessary to meet overall performance objectives. Automated control systems that modulate blower output based on process parameters can be implemented with relatively low capital investment and provide significant energy savings compared to constant blower output or manually adjusted systems. The more sophisticated aeration control systems, which incorporate actuated flow control valves, add a further layer of complexity to the control system programming and capital expense, but such advanced controls might be warranted for facilities that require more precise control to optimize performance of a BNR process. The latter is often driven by stringent effluent quality requirements, which is an important non-cost factor that can drive a facility toward implementing advanced aeration controls beyond simply the cost–benefit around power cost savings. Refer to Section 2.3 for further discussion of control strategies that can be implemented to minimize pressure losses induced by automated control valves.

The categories of aeration control discussed in this section include automated controls based on process parameters (dissolved oxygen, ammonia, and nitrate) and how those parameters monitored in real time can be used to modulate blower output and/or airflow control valves within the aeration system piping.

4.2 Constant Aeration Rate

Operating the aeration system at a constant airflow rate with no adjustment based on measured process parameters or influent loading characteristics is the simplest means to control airflow. However, it is also the most energy-intensive method because the system must be operated at higher airflow rates than strictly required to meet the process requirements. Additionally, this method is more prone to "over-aeration" of the mixed liquor, which can promote foaming and other process performance issues.

4.3 Parameter-Based Control

4.3.1 Dissolved Oxygen

The most common parameter used for activated sludge aeration systems is dissolved oxygen concentration measured in the mixed liquor aeration zones. This section discusses the various ways that dissolved oxygen probe information can be used to control aeration system output.

4.3.1.1 Intermittent Monitoring and Manual Aeration Adjustment

This method of blower control relies on dissolved oxygen readings either from a hand-held probe or from a permanently installed probe that operators monitor. Either way, operators choose to make manual adjustments to the blower system's output using their judgment on the measured versus target dissolved oxygen concentration and the time of day relative to the WRRF's typical diurnal loading requirements. In this case, the dissolved oxygen concentration data are used purely for monitoring purposes, not for automated control. This approach might be suitable for smaller WRRFs, where overall energy costs are relatively low. However, once the investment is made in permanently installed dissolved oxygen probes within the tanks, the additional cost to provide some level of automated dissolved oxygen control will result in a favorable energy savings and a cost payback for most WRRFs.

4.3.1.2 Automation with Constant Dissolved Oxygen Setpoints

Automated dissolved oxygen control based on an operator-entered dissolved oxygen concentration setpoint is the most commonly used approach for automated aeration control at WRRFs. Automated dissolved oxygen control systems without flow control devices in the basin are typically designed as a feedback loop, in which the blower is modulated in response to the variance between the measured dissolved oxygen concentration and the target dissolved oxygen concentration setpoint.

The dissolved oxygen variable can rely simply on a single dissolved oxygen measurement in one tank or an average of dissolved oxygen probes across multiple tanks. Facilities with flow control valves on each drop leg may also provide multiple dissolved oxygen probes in each tank to target different dissolved oxygen setpoints within the basins, depending on the process requirements. Historically, the conventional wisdom was to target a concentration of 2 to 3 mg/L dissolved oxygen to ensure sufficient aeration for all loading conditions. However, as discussed in Section 3.3, more recent process control philosophies promote the use of lower dissolved oxygen setpoints to provide conditions to accomplish SND, nitrite shunt, and other low-energy nitrogen removal processes.

It is important to note that placement of the dissolved oxygen probe in a plug flow aeration basin will have a significant effect on the overall energy use, and on the overall performance of the process. Placement of a single dissolved oxygen probe at the beginning of the basin to meet a setpoint of 1 to 2 mg/L dissolved oxygen will result in elevated dissolved oxygen concentrations at the end of the basin, which could interfere with efficient denitrification. Placement of a single dissolved oxygen probe at the end of the basin to meet a setpoint of 1 to 2 mg/L dissolved oxygen will result in dissolved oxygen concentrations below 1.0 mg/L at the beginning of the basin, which has the potential to cause unwanted operational problems, such as foaming. Selecting the appropriate number of dissolved oxygen probes and optimal probe locations should consider the treatment objectives and effluent requirements, and then consider the appropriate setpoint given the probe location. The maintenance and calibration of the dissolved oxygen probes should also be considered in selecting the number of probes required to achieve process control.

4.3.1.3 Automation with Dynamic Dissolved Oxygen Setpoints

An additional step of automation complexity can be provided in which dissolved oxygen setpoints are calculated values based on the measurements of an alternate controlled variable, such as ammonium. Many ABAC systems (though not all) use ammonium concentration as the primary controlled variable, which is then used to calculate a dynamic, variable dissolved oxygen setpoint (or multiple dissolved oxygen setpoints) to maintain the desired ammonium concentration.

4.3.2 Ammonia-Based Aeration Control

Although the concept is not new, ABAC has recently seen increased application at municipal WRRFs resulting from improvements in ammonia instruments and increased focus on energy and process control. Following are the likely energy and operational benefits that WRRFs can achieve using a well-implemented ABAC system:

- Reduced blower power consumption made possible by limiting nitrification (assuming the blowers have sufficient turndown capability);
- Reduced supplemental carbon costs for BNR treatment, which can be made possible through several mechanisms: (1) Limiting nitrification results in less nitrate that must be denitrified and (2) If dissolved oxygen is allowed to decrease, then simultaneous nitrification/denitrification might be possible;
- Reduced supplemental ammonia costs (for facilities that use chloramine disinfection);
- Faster aeration system response to influent ammonia spikes (i.e., higher aeration rates in response to higher influent loading);
- Lower shear on the biofloc and improved settling characteristics, presumably because of lower airflows; and
- Improved detection of upsets caused by influent toxic constituents and overall improved monitoring of nitrification performance (because ammonia is directly monitored rather than relying on dissolved oxygen as a surrogate for nitrification).

Some of these benefits are attractive given the potential for reduced operational costs. However, implementing an ABAC system does incur additional cost to purchase and properly maintain the ammonium instruments and to keep the control system functioning optimally.

Ammonia-based aeration control systems can be designed in a variety of ways. One consideration is the use of feedback versus feedforward control strategies. Feedback strategies are simpler to implement and require fewer instruments, but they also introduce a longer lag time because the change in loading at the front of the tank takes time to reflect a change in concentration at the end of the tank. Feedforward strategies can provide faster response times but also require more instruments, and in many cases contain control logic that the vendor considers to be proprietary. Another consideration is whether an ABAC system will cascade its output into a dissolved oxygen control system, versus produce an output that directly outputs calculated flowrates to modulate the blower (or control valves if those are present). The cascaded approach in which the ABAC controller calculates variable dissolved oxygen setpoints that are used in a dissolved oxygen control system is most common in the United States, but a few WRRFs have implemented the direct ABAC approach because of its simplicity and reportedly faster response time; however, pilot testing of this method is recommended to evaluate appropriate fail-safe alarms and controls are in place. Regardless of the control strategy used, maintaining some form of dissolved oxygen measurement and high/low dissolved oxygen alarms and limits is recommended.

4.3.3 Ammonia-vs NOx-N (AvN™) Control

Ammonia-vs NOx (AvN™) is a patented control scheme that aims to achieve nitrite shunt in mainstream treatment processes, which is a less energy-intensive means to accomplish nitrogen removal compared to conventional nitrification/denitrification processes. The goal is to minimize total inorganic nitrogen by promoting ammonia-oxidizing bacteria when measured ammonia exceed the measured nitrate/nitrite concentration and then promoting nitrite-oxidizing bacteria when the measured nitrate/nitrite concentration exceeds the measured ammonia. Thus, this controller relies on dissolved oxygen, nitrate, and ammonia concentrations as input, and the resultant output is to control the blowers in an intermittent (but automated) fashion to produce equal effluent concentrations of ammonia and nitrate/nitrite. The AvN™ controller has been used at several WRRFs worldwide.

4.4 System Air Flow Control

The output from the process parameter control loops discussed in Section 4.3 is typically fed into a flow control loop to determine how much airflow is required to maintain the process setpoint. This means the parameter control loop calculates the airflow required for a given control zone based on the difference between the process setpoint and the actual reading from the real-time analyzer. Smaller facilities typically do not have multiple control zones, so the change in airflow required is used directly to increase or decrease the total system flowrate from the blowers. The distribution of flow to individual bioreactors or zones within a bioreactor is done manually. For more advanced control systems, multiple control zones are created by installing flow meters and flow control valves to automatically distribute airflow. In these systems, the valve position for the associated control zone is adjusted to meet the airflow setpoint calculated by the parameter control loop. Adjustment of the total system flowrate from the blowers is then done using either pressure- or flow-based controllers. These controllers are typically paired with a most open valve (MOV) scheme for the control valves to further reduce energy consumption. Most open valve control is discussed in Section 4.5.

4.4.1 Constant Pressure Control

Constant pressure system operation provides a set discharge pressure that the blower equipment is required to maintain. Jenkins (2013) notes that pressure was added as a control variable for the main system header to overcome the loop independence issues created by the inability of older controllers to share data in real time. In a typical older system, all loops were essentially independent of each other even if there was a process relationship between them. However, maintaining constant pressure has a negative effect on system power requirements because it does not allow the pressure to drop as the airflow requirement decreases. It can also create instability and hunting in the system. This strategy is essentially obsolete because of this.

4.4.2 Variable Pressure Control

Variable pressure control systems were developed to minimize unnecessary pressure losses inherent to constant-pressure aeration systems and, therefore, reduce energy consumption. Variable pressure system operation is similar to constant-pressure strategy, except it does not have a single defined pressure setpoint to maintain. The pressure setpoint is calculated based on the position of the most open control valve, which allows the system to operate at a lower overall pressure and to save energy. With a variable pressure system, the system pressure will increase or decrease as the parameter control system fluctuates. For example, if less airflow is required because the process parameter is above the setpoint, there will be a corresponding decrease in system pressure.

Variable pressure control systems became widely used in the 2000s and were typically programmed using cascading proportional-integral-derivative (PID) loops. Proportional-integral-derivative loops are commonly used for analog control and can be an excellent control strategy for many applications. Aeration control with PID loops can be challenging, however, because PID loops are ideally suited for linear response systems and most aspects of aeration systems are nonlinear. The complexity increases for dissolved oxygen-based control systems because the relationship between dissolved oxygen and airflow is dependent on a variety of factors (including temperature, oxygen mass transfer, and other process-related factors) that are constantly changing and dissolved oxygen control has significant process time lags. This makes it difficult to provide a well-tuned and stable control system.

In recent years, variable pressure control with non-PID based loops has been developed as an alternative to PID systems. These systems use a "proportional and pause" approach, not PIDs. For a dissolved oxygen loop, changes in airflow required to meet the dissolved oxygen setpoint are adjusted in direct proportion to the error and then the system pauses before making adjustments. The "pause" part of this approach helps to prevent cycling, which is a common undesirable behavior in pure PID feedback loops.

4.4.3 System-Level Flow Control

Direct flow control is an alternative to pressure control. Some implementations calculate the total blower flow setpoint as an absolute value, meaning that the setpoint is taken as the sum of the individual control zone air demands. Others adopt a proportional approach to determining the blower flow setpoint, such that the blower flow setpoint is only changed in response to the sum of the changes (or error) in airflow at each control zone. An advantage of this error-based flow control approach is that it accounts for any unmeasured flows in the system (e.g., leakage) because the amount of air to be allocated to that demand does not need to be known.

System-level flow control is best implemented without PIDs, using a "proportional and pause" approach instead. Some refer to this as an "adjust-wait-and-see" approach, and Jenkins (2013) refers to it as "floating control." A good analogy for this type of control is someone adjusting water temperature while taking a shower: he/she checks the temperature, taps the hot water valve open or closed a little, and then waits for the temperature to stabilize. The process is repeated until the shower temperature is within the comfort zone (Jenkins, 2013). This type of control can be effective for aeration systems because of their nonlinear nature, as long as the incremental change is small. The other benefit of adding the "pause" is to avoid amplification or propagation of feedback oscillations from the parameter and control zone airflow control loops through the system-level flow control system. The "pause" serves to dampen these oscillations out rather than immediately react to them.

4.5 Most Open Valve Control

Facilities implementing automated modulating valves must determine how those valves will be regulated in the automated control system. Objectives for the aeration control system, of which the control strategy is an important element, should include

- Achieve the desired process parameter setpoint (often dissolved oxygen) quickly and maintain that setpoint under variable loading conditions;
- Maintain the desired process setpoints with as few actuator starts/stops as possible to minimize actuator maintenance; and
- Optimize energy usage by minimizing pressure loss in the system.

Many are familiar with the term "most open valve" control, but the exact meaning of the term is not always clear. An MOV algorithm with header pressure control was developed as far back as 1978 and was used to optimize efficiency in steam systems with a common header. Throttling is an irreversible process that wastes energy, and to minimize those losses, it is desirable to have control valves further open while still maintaining control of their individual variables (Cho, 1984). This is the same basic principle modern aeration control MOV systems are trying to satisfy: maintain at least one airflow control valve at its maximum position at all times. Maximizing valve opening reduces pressure loss across the valves, reducing system pressure and therefore power demand.

Most open valve control is a secondary or cascade control strategy in aeration control; it is not part of maintaining the process parameter setpoint, nor is it part of the blower system airflow controls. Most open valve control is used when there are multiple airflow control valves in the aeration control system, which are typically at different positions. The MOV position setpoint is determined in a way to maintain the position as open as possible within an acceptable range for the type of valve installed in the system.

The earliest iterations of MOV control were developed as an improvement upon control systems that used modulating valves to balance airflows within the system, and yet the blowers were set to maintain a constant pressure setpoint regardless of the valve position downstream. Variable pressure based MOV systems were designed to address this inefficient constant pressure control scheme. Since the advent of variable pressure based MOV control systems, other types of MOV control strategies have been developed, namely direct flow control MOV. Figure 9.3 presents a variable pressure MOV scheme and Figure 9.4 presents a flow MOV scheme for system-level control.

When MOV is used in conjunction with variable pressure system control, three cascaded control loops are involved: (1) the parameter-based controller calculates an airflow setpoint for each control zone, (2) the flow/valve controller calculates a valve position setpoint to meet the airflow setpoint calculated by the first loop, and (3) the blower pressure controller modifies the blower system operation to maintain a floating pressure setpoint calculated based on the MOV position. The floating pressure setpoint calculation is characteristic of the individual system. In many systems, the controller will increment or decrement the pressure setpoint if the MOV is above or below preset limiting values. In other systems, the floating pressure setpoint calculation has a minimum pressure plus a factor multiplied by valve position of the position of the MOV. This means the pressure setpoint will increase as the MOV approaches the maximum allowable position because more air is required in the system. The increase in header pressure setpoint results in an increase in blower system airflow. If the MOV reaches the minimum allowable position, the pressure setpoint will be reduced because less air is required to meet the parameter setpoint. This results in a decrease in blower system airflow and pressure, saving energy in periods of low demand. Variable pressure MOV does not function properly if blowers are at minimum or maximum capacity because changing the pressure setpoint cannot change the blower flowrate. This results in flow control valves moving to extreme positions that can result

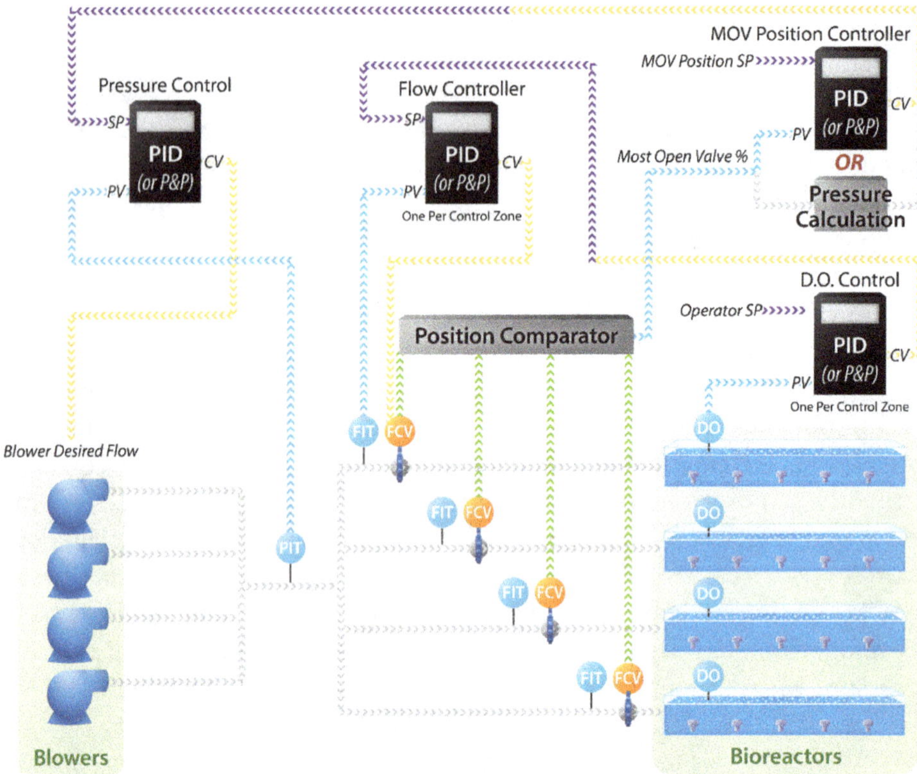

FIGURE 9.3. Variable pressure MOV schematic. Courtesy of CDM Smith.

in reduced accuracy and increased system pressure and energy consumption if the valves are throttled to minimum positions.

When MOV is used in conjunction with flow system control instead of a pressure control, the control scheme incorporates the same parameter-based controller and flow/valve controller as a variable pressure system. The pressure controller is eliminated, and the blower system output is adjusted to maintain the calculated system airflow setpoint while the valve positions are modulating. As explained in Figure 9.5, valves modulate to maintain airflow as a proportion (percentage) of total airflow available. A close permissive must be programmed for each control zone to determine if a valve can be closed or not in response to the airflow set point. The MOV will not be allowed to close even if the airflow in that control zone is above the setpoint. If the airflow is higher than the setpoint in one control zone, it will necessarily be lower in another control zone, and the control valve for that zone will cycle open. This will happen for a few iterations and the flows will either be correct, or a different valve will reach the most open position. If a different valve becomes the most open, the previous most

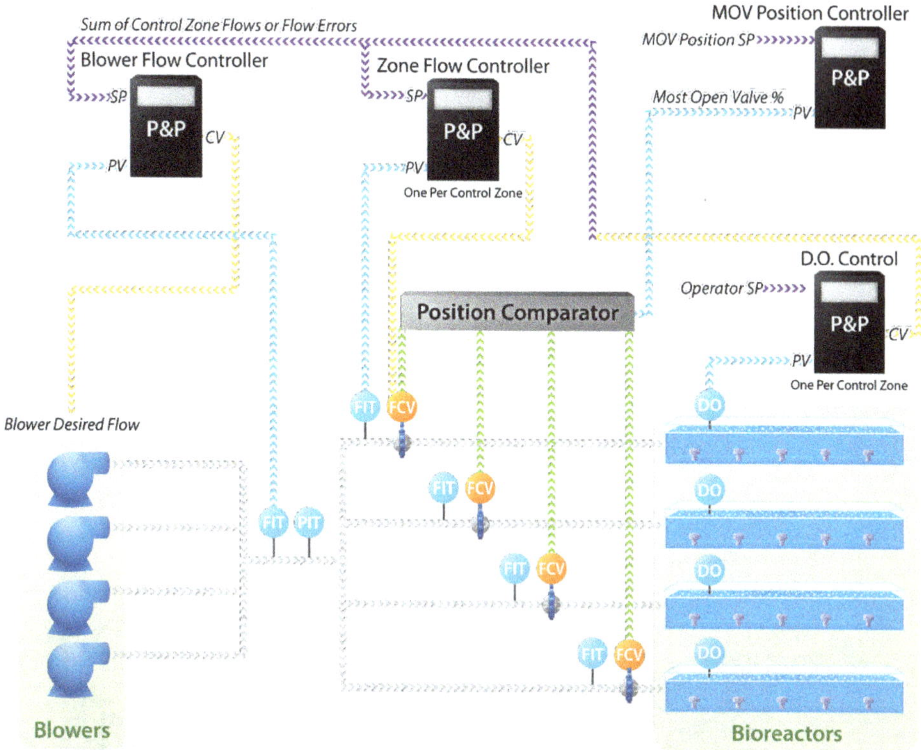

FIGURE 9.4. Flow MOV schematic. Courtesy of CDM Smith.

open valve will be allowed to close to reduce airflow to the control zone and achieve its setpoint.

Table 9.2 presents a summary of variable pressure and flow MOV control schemes as well as advantages and disadvantages of each. Flow-based MOV offers some energy savings over variable-pressure MOV because the blowers are responding directly to a flow setpoint and because the one flow control valve is always at its maximum position rather than between positions.

Achieving good process control by implementing an MOV scheme will improve energy efficiency at a WRRF by lowering the system pressure in the header. Lowering the system pressure reduces the energy consumption of the blowers, which typically account for the highest energy demand. Non-PID MOV control offers additional benefits from a system stability standpoint, which could improve biological process performance. Flow-based non-PID MOV offers the potential for the most energy savings because it removes pressure as a control variable and because one flow control valve is always at its maximum position rather than between positions.

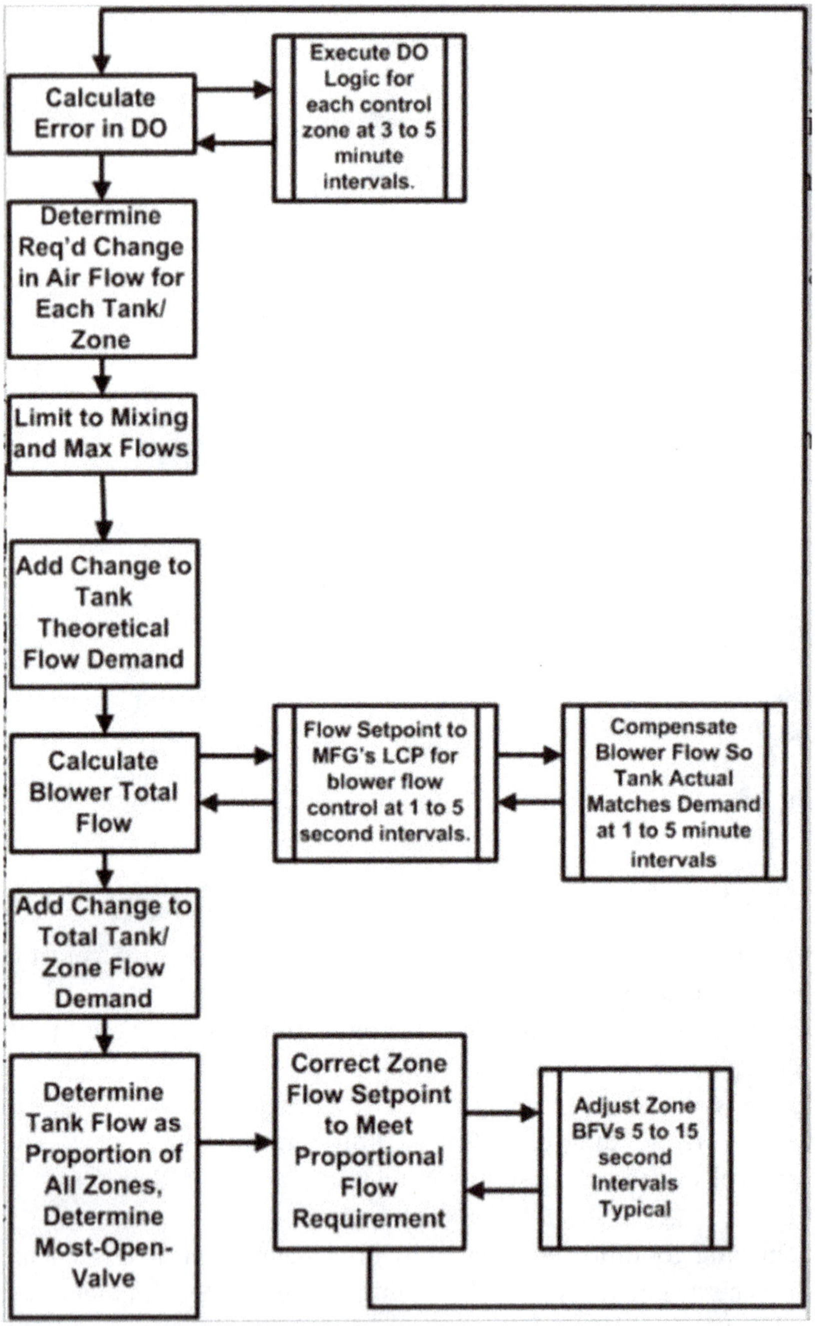

FIGURE 9.5. Flow MOV control logic description

TABLE 9.2. MOV programming alternatives for parameter-based control (adapted from Neville et al., 2019)

Type	Description	Advantages	Disadvantages
Variable Pressure PID Blower Control with Dissolved Oxygen PID Loops	• Three cascades of PID control loops • Dissolved oxygen PID loop assumes linear relationship between airflow and dissolved oxygen. • Linear relationship is established during system startup/tuning. • Blower responds to change in header pressure setpoint, which is determined based on position of MOV.	• Well-established • More widely known by system integrators • Pressure transmitter provides a single, stable reading for blower control.	• Fixed adjustment of pressure is more susceptible to oscillation/hunting in response to major system changes. • Increases response time of control system • Does not function properly if blowers are at minimum or maximum capacity. • Re-tuning dissolved oxygen and pressure PIDs may be required more often for optimal performance because relationships between controlled variables change depending on ambient conditions and number of operational blowers.
Variable Pressure Non-PID Blower Control with Floating Dissolved Oxygen Loops	• Dissolved oxygen and valve control loops use a "Proportional and Pause" approach; not PIDs. For dissolved oxygen loop, changes in airflow required to meet dissolved oxygen setpoint are adjusted in direct proportion to the error and then system pauses before adjusting. • Blower pressure control loop uses fixed adjustments rather than proportional adjustments. • Blower responds to change in header pressure setpoint, which is determined based on position of MOV.	• Well-established • Less widely known by system integrators than systems using PIDs • Pressure transmitter provides a single, stable reading for blower control. • Minimizes oscillations because dissolved oxygen PID loops are not used	• Fixed adjustment of pressure is more susceptible to oscillation/hunting in response to major system changes. • Does not function properly if blowers are at minimum or maximum capacity. • Re-tuning may be required periodically for optimal performance, although likely less necessary than for PID-based. • May require process expertise rather than PID loop knowledge to tune.

(*continued*)

TABLE 9.2. *(Continued)*

Type	Description	Advantages	Disadvantages
Flow Non-PID Blower Control with Floating Dissolved Oxygen Loops	• All control loops (dissolved oxygen, valves, and blowers) use a "Proportional and Pause" approach, not PIDs. For dissolved oxygen loop, changes in airflow required to meet dissolved oxygen setpoint are adjusted in direct proportion to the error and then system pauses before adjusting. • Blower responds to airflow setpoint and is adjusted simultaneously with valve position.	• Blower output matches system demand. • Minimizes oscillations because dissolved oxygen PID loops are not used • Might result in more stable operation and/ or faster response time, particularly for systems with long piping runs • Might result in a lower header system pressure and energy savings because one flow control valve always in most open position rather than between positions	• Flow signals are less stable than pressure readings, but signals can be dampened or change in flow (delta) can be used instead. • Re-tuning may be required periodically (seasonally) for optimal performance, although likely less necessary than for PID-based. • May require process expertise rather than PID loop knowledge to tune.

4.6 Blower Output Control

Either pressure or airflow is used to determine what the blower system must output to meet process demands, but the way in which blower output is modulated is different depending on the type of blower installed. Chapter 8 presents the control techniques available for different types of blowers.

5.0 ULTRAVIOLET DISINFECTION

5.1 Introduction

Ultraviolet disinfection is in common use at WRRFs. The appeal of ultraviolet disinfection is that it requires no chemical injection into the facility's effluent. It also has a small process footprint compared to other disinfection alternatives, making it a feasible option. However, ultraviolet disinfection systems consume more energy than other commonly used disinfection systems (e.g., chlorination). Most ultraviolet systems are designed to achieve adequate disinfection under the most challenging conditions of flow and effluent quality that are anticipated at the WRRF; however, most of the time the ultraviolet system will not be subjected to such conditions. As a result, there is an opportunity to save energy during typical operating conditions, while still maintaining adequate ultraviolet dose, with the application of adequate monitoring and controls.

5.2 Data Collection and Monitoring

Successful operation of an ultraviolet disinfection system requires maintaining an adequate ultraviolet dose to inactivate the target organism and meet the WRRF's effluent disinfection limits. The following data is recommended to properly monitor ultraviolet system operation and maintain dose:

- **Flowrate.** Continuous measurement of flowrate through the ultraviolet reactor is necessary to calculate ultraviolet dose delivered, and to the implementation of automatic system control (see below).
- **Effluent bacteria count.** Regular monitoring of ultraviolet reactor effluent for the target bacteria count is required for most National Pollutant Discharge Elimination System (NPDES) permit reporting. Review of the data can also indicate whether it is appropriate to modify the delivered ultraviolet dose to maintain compliance while also reducing energy use.
- **Ultraviolet intensity (UVI).** Ultraviolet intensity is a measure of the actual level of ultraviolet radiation in the bioreactor and is necessary to calculate ultraviolet dose delivered. Ultraviolet intensity sensors are typically provided by the ultraviolet system manufacturer. Care should be taken in interpreting UVI data, because the UVI sensor is typically located adjacent to a subset of ultraviolet lamps and may not be representative of the entire ultraviolet reactor if the lamps and sleeves in the system differ in terms of age, performance or cleanliness.
- **Ultraviolet transmissivity (UVT).** Ultraviolet transmissivity is a measure of how well ultraviolet radiation passes through the water being disinfected, such that poorer-quality water will have a lower UVT and high-quality water will have a higher UVT. The UVT greatly affects the energy needed to deliver the required ultraviolet dose. Ultraviolet transmissivity is a primary indicator of the anticipated performance of the ultraviolet system and is often an input into automatic system control (see below). In situ, continuous UVT sensors are typically provided by the ultraviolet system manufacturer as part of the overall installation but can also be procured separately. Hand-held UVT sensors are also available.
- **Ultraviolet disinfection influent bacteria count.** Though not routinely monitored, the influent bacteria count to ultraviolet systems is fundamental as to the log inactivation required by the system, and

therefore the ultraviolet dose that is necessary to successfully meet permit. Because this data is commonly lacking, ultraviolet systems are typically operated at a sufficiently large dose (perhaps overdose) to ensure permit compliance.

5.3 Control and Energy Efficiency Concepts

As stated above, ultraviolet disinfection systems are sized to deliver the necessary dose at the WRRF's peak flowrate, and a conservative (if not worst-case) UVT. Thus, most of the time the output of the ultraviolet system can be operated at less than capacity and still meet disinfection requirements. Smaller WRRFs may find it preferable to simply run the ultraviolet system at 100% (i.e., overdose) at all times to ensure adequate disinfection; however, at larger WRRFs, or in locations where energy efficiency is a primary goal, automatic operation can save energy. Automatic control approaches, in varying leveling of complexity, can include

- **Flow pacing.** Simply flow pacing the ultraviolet system output results in significant reduction in energy use compared to operating the system at 100% output. This approach does not consider that the water quality, as monitored by UVT, varies.
- **Fine-tuning based on UVT.** Increased complexity of ultraviolet system operation can be achieved by incorporating continuous UVT measurement into the system controls. In such systems, the system PLC adjusts ultraviolet lamp output downward if UVT is high, and increases output when UVT is low, to maintain the dose setpoint.

Although not strictly automatic controls-based, the energy consumption of an ultraviolet disinfection system can be greatly reduced by regularly cleaning the ultraviolet reactor and the quartz sleeves. Algae, foam, debris, and settled solids particles can both increase the bacteria counts that must be disinfected and decrease the efficiency of ultraviolet transmission. The required ultraviolet dose can also be reduced by optimizing the process performance of upstream processes, which increases the UVT of the ultraviolet system influent. At WRRFs that dose iron compounds in upstream processes for enhanced settling or for phosphorus removal, it is important to maintain tight control of iron dose, as residual iron in the ultraviolet system influent will absorb ultraviolet light and thereby reduce UVT.

6.0 REFERENCES

Cho, C. H. (1984). *Computer-Based energy management systems: Technology and applications* (1st ed.) Academic Press.

Doody, A. T., & Neville, M. D. (2017). Design and operation of advanced aeration control systems. *Proceedings of the 90th Annual Water Environment Federation Technical Exhibition and Conference;* Chicago, Illinois, Sept 30–Oct 4. Water Environment Federation.

Jenkins, T. E. (2013). *Aeration control system design: A practical guide to energy and process optimization* (1st ed.). John Wiley & Sons, Inc.

Lipták, B. G., & Venczel, K. (2016). *Instrument and automation engineers' handbook: Process measurement and analysis* (5th ed.). Taylor & Francis.

Neville, M. D., Doody, A. T., Hussain, S., Sangrey, K., Brown, J., & Hilgart, T. (2019). New aeration controls for improved BNR performance and cost savings. *Proceedings of the 92nd Annual Water Environment Federation Technical Exhibition and Conference;* Chicago, Illinois, Sept 21–25. Water Environment Federation.

Water Environment Federation (2013a). *Automation of water resource recovery facilities* (4th ed.; WEF Manual of Practice No. 21).

Water Environment Federation. (2013b). *Operation of nutrient removal facilities* (WEF Manual of Practice No. 37).

7.0 SUGGESTED READINGS

Rieger, L., Jones, R. M., Dold, P., & Bott, C. B. (2014). Ammonia-based feedforward and feedback aeration control in activated sludge processes. *Water Environment Research, 86*(1).

U.S. Environmental Protection Agency. (2010). Design and control of aeration systems. In *Evaluation of Energy Conservation Measures for Wastewater Treatment Facilities* (EPA 832-R-10-005). U.S. Environmental Protection Agency Office of Wastewater Management.

10

Digester Gas to Energy Systems

Robert (Bob) Forbes, PE; Xiaojin Li, PhD; Shayla Allen; Anthony Tartaglione, PE, BCEE; Jim Groman; Houssam El Jerdi, MS, PE; Jay Kemp; & Eric Auerbach

1.0 INTRODUCTION

The purpose of this chapter is to describe how digester gas, which is produced from the anaerobic digestion process at water resource recovery facilities (WRRFs), is produced, captured, treated, and used to recover energy.

1.1 Chapter Objectives

The objectives of this chapter are

- to explain how digester gas is produced and enhanced by the anaerobic digestion process;
- to describe the characteristics of digester gas, focusing on its value as biofuel for energy production;
- to explain how digester gas is captured, handled, treated (conditioned), and recovered to produce energy; and
- to introduce and describe the different methods of recovering digester gas for energy.

1.2 Definition of Digester Gas

Biogas is generally defined as a group of gases produced by the degradation of organic material in the absence of oxygen (also known as anaerobic degradation) and dominated by methane (CH_4), which is the primary component of biogas that can be burned to produce energy. *Natural gas* is typically defined as fossil biogas that was produced by anaerobic degradation eons ago and has been stored deep in the earth's rock formations. Biogas is produced by the same types of anaerobic degradation processes that produce natural gas, except that the anaerobic degradation and biogas production occur in the present.

Digester gas is a type of biogas that is specifically the product of anaerobic digestion of wastewater residuals and other feedstocks at a WRRF. There are other types of biogas produced by anaerobic processes in solid-waste landfills, some industrial processes, and other processes at WRRFs (such as upflow anaerobic sludge blankets); however, those other sources of biogas may differ from digester gas in several respects, and they are not covered in this chapter.

Except for limited discussion on blending *digester gas* with other forms of *biogas* or *natural gas* to increase the overall energy potential of biogas, this chapter focuses strictly on the characterization, production, conditioning, and utilization of *digester gas* from WRRFs.

2.0 DIGESTER GAS PRODUCTION

This section describes the mechanisms and rates by which anaerobic digestion of biosolids forms digester gas.

2.1 Anaerobic Digestion

Digester gas from a WRRF evolves from a chain of biochemical reactions in the absence of free oxygen, referred to as anaerobic digestion, in which proteins, fats, and carbohydrates contained in wastewater treatment residuals (including primary and secondary sludges, scum, and imported organic feedstocks) are treated in purposed-built reactors known as *digesters*. The feedstocks are degraded in digesters through several steps and pathways into biosolids and gases such as, ammonia, hydrogen, water vapor, carbon dioxide (CO_2), and CH_4, the primary fuel component of digester gas. The goal of anaerobic digestion from an energy perspective is to maximize the production of digester gas and specifically its CH_4 content, which is the primary fuel in digester gas.

Biochemical pathways of anaerobic digestion that form digester gas are shown in Figure 10.1.

2.1.1 Microbial Community

Digester gas is produced via synergistic interactions between a consortium of anaerobic bacteria and archaea. In the anaerobic digestion process, redox reactions can occur in series and in parallel to convert organic matter to the most reduced forms of carbon, primarily CH_4 and CO_2. The CH_4 is created

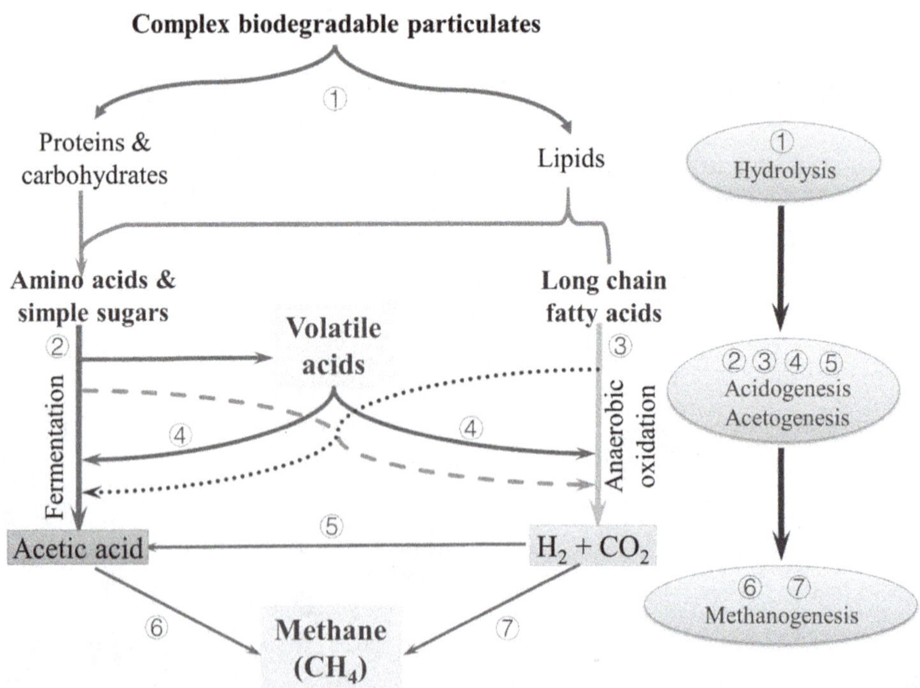

FIGURE 10.1. Anaerobic digestion biochemical pathways

primarily from a biochemical process called methanogenesis. While other biochemical processes such as sulfidogenesis also occur, detailed evaluation of these processes is out of scope for this chapter that focuses on the energy potential of digester gas.

Understanding of microbial communities and their functions is important to improve the efficiency and process stability of anaerobic digesters, and to maximize the production of digester gas. The digestion process is described in full in other references, such as the *Solids Process Design and Management Manual* (Water Environment Federation [WEF] et al., 2012) and the *Design of Water Resource Recovery Facilities* (WEF et al., 2018).

2.1.2 Digestion Inhibition and Toxicity

Toxicants may either be present in the feedstock or produced during digestion; in either case, toxicity will severely curtail or shut down the production of methane. Nutrients and trace elements can act as toxicants depending on their chemical form and dissolved concentrations. Many studies have suggested that inorganic compounds, such as free ammonia, heavy metals, and sulfide can inhibit the digestion performance. Some studies found that adding certain trace elements to anaerobic digesters significantly improved operation stability and performance under high ammonia concentration (Banks et al., 2012; Bayr et al., 2012). However, the overall energy balance and economic benefits should be taken into consideration before implementing these technologies.

2.1.3 Volatile Solids Reduction

Volatile solids (VS) and chemical oxygen demand (COD) are common measures of the organic substrate entering a digester. Volatile solids are considered to represent the organic fraction of biosolids, part of which is biodegradable and can be converted to digester gas. Volatile solids measurements are typically used as part of determining overall process performance, regulatory compliance, and for mass-based calculations.

One must be careful when using data from a VS test, because VS results do not always represent the organic fraction accurately. A significant amount of inorganic salts—especially ammonium-based salts—can volatilize in analytical tests, skewing the VS concentration higher and the VS destruction lower (Beall et al., 1998).

2.1.4 Other Process Indicators

Chemical oxygen demand is a measure of the chemically oxidizable material in solids. As with VS, this measure overestimates the biodegradable fraction, but process modeling can compensate for this error. Chemical oxygen demand typically is used in kinetic process models because the model parameters are based on COD coefficients (Batstone et al., 2002; WEF, et al. 2018). Most WRRFs report VS data rather than COD data because the VS test is quicker and more economical than COD analysis. It is useful to develop correlation curves of COD to VS for each anaerobic digestion system so that estimates of COD can be made based on measurements of VS, and ratios of COD to VS are largely consistent for each anaerobic digestion system.

There are many details to consider in optimizing anaerobic digestion systems, primarily in maintaining recommended design and process parameters such as hydraulic residence time (HRT), solids residence time (SRT), design temperature (e.g., mesophilic or thermophilic temperature ranges), VS loading rates, and mixing energy/type. The anaerobic digestion process must be designed and operated to maintain all process parameters within recommended design standards, the details of which are covered in other references (WEF et al., 2012; WEF et al., 2018). Optimizing the anaerobic digestion process through proper design and operation results in maximizing the production and quality of digester gas.

The remainder of this chapter focuses on the characterization, generation, treatment, and beneficial uses of digester gas from the anaerobic digestion process to maximize recovery of its energy potential.

2.2 Methods to Estimate and Enhance Digester Gas Production

This subsection describes how digester gas production is estimated and includes some of the methods by which anaerobic digestion can be made more effective and efficient, thereby increasing digester gas production.

2.2.1 Digester Gas Production and Estimation

As shown in Figure 10.1, digester gas is formed mostly during the last stage of digestion, when microorganisms convert volatile fatty acids (VFAs) to CH_4, CO_2, and water vapor, among other gases in lower concentrations. Digester gas production is directly related to the amount of VS destroyed and is typically expressed as volume of gas per unit mass of VS destroyed, also known as VS reduction (VSR). Digester gas production rates are different for each organic substrate in the digester. The unit production and CH_4 content of digester gas resulting from the degradation of various sludge constituents are presented in Table 10.1.

Specific gas production for each anaerobic digestion system should be measured until an average unit value of gas yield can be obtained and used for monitoring for that specific digestion system.

The two main constituents of digester gas are CH_4 and CO_2; it also contains nitrogen, hydrogen, and hydrogen sulfide (H_2S), among other trace gases. Performance data from healthy anerobic-digestion systems suggest

TABLE 10.1. Digester gas unit production and methane (CH_4) content based on substrate (Shen et al., 2015; WEF et al., 2018)

Anaerobic Digestion Substrate	Gas Yield per Unit VSR, m^3/kg (cu ft/lb)	Range of CH_4 Content (%)
Fat	1.2–1.6 (19–26)	62–72
Scum	0.9–1.0 (14.4–16)	70–75
Grease	1.1 (17.6)	68
Crude fibers	0.8 (12.8)	45–50
Protein	0.7 (11.2)	73
Primary sludge (PS)	0.9–1.2 (14.4–19)	65–70
Waste activated sludge (WAS)	0.7–0.8 (11.1–12.8)	60–65
PS+WAS blend (typical)	0.8–1.1 (12.8–17.6)	60–70

that CH_4 concentrations in digester gas should be 60% to 70% and CO_2 concentrations should be 30% to 35% (by volume). Methane concentrations under 60% in digester gas may indicate excessive air intrusion or contamination from an outside source, while CO_2 levels above 35% in digester gas often indicate an upset digester. Concentrations of H_2S (above 1% by volume) can indicate an unbalanced digestion process, industrial waste sources, or saltwater infiltration. Hydrogen sulfide is the primary compound responsible for odors and corrosion in digester-gas systems. Some metals such as aluminum and iron can precipitate as metallic sulfides, thereby minimizing H_2S concentrations in digester gas. Other effects of H_2S on digester gas recovery are described in Section 3.0.

Digester gas production rates should be routinely monitored to determine the average gas production for a corresponding VSR. Anaerobic digestion processes that are well designed and operated to capture digester gas efficiently should produce digester gas at the higher ends of the gas yields per unit VSR shown in Table 10.1. If digester gas production is found to be at the lower ends of the gas-yield ranges shown in Table 10.1, then it is recommended to more closely examine the condition and operation of digesters and gas management systems.

2.2.2 Advanced Digestion Processes

Advanced digestion processes involve modifications to the conventional mesophilic digestion to achieve better VSR and digester-gas production, enhance pathogen reduction, and improve digester operation (Forbes, 2013).

Advanced digestion processes may also increase the content of soluble nutrients (ammonia and phosphorus) in dewatering side streams, and thus can increase the nutrient loads to the wastewater treatment system. Whole-facility process modeling should be used to predict the effects of modifying or enhancing the anaerobic digestion process at a WRRF. See Chapter 4, Section 2.0—Use of Process Models for a more complete discussion of the role of whole-facility process models in WRRF processes and energy management.

2.2.2.1 Two-Phase Anaerobic Digestion

Two-phase digestion or acid/gas digestion is carried out in a two-reactor system to provide separate environment for the acid-forming and methane-forming microorganisms, so each reactor can be optimized for the specific conditions. In the acid phase, the feed substrates are hydrolyzed to produce VFAs, which are then converted to CO_2 and CH_4 in the gas phase. Typically, the acid phase is limited to an SRT of 1.5 to 2 days, while the gas phase requires a minimum SRT of 10 to 15 days because of the low growth rate of methanogens. The acid phase produces gas that is typically flared because of its lower quality and higher concentrations of contaminants such as H_2S and siloxanes. Subsequently, the gas phase typically produces digester gas of higher quality and energy value than the digester gas from conventional anaerobic digestion processes.

2.2.2.2 Thermophilic Anaerobic Digestion

Thermophilic digestion processes include one or more stages that are operated in the thermophilic temperature range, which is typically between 50 and 57 °C (122 and 135 °F). The goals of thermophilic digestion are to achieve greater pathogen destruction, increase VS destruction, and reduce detention time. Disadvantages of thermophilic digestion can include higher digester heating costs, potential reduction in dewatering capability, and increased odors during dewatering. Heat recovery is an important step to reduce heating costs in thermophilic digestion. Waste heat from co-generation systems as well as thermophilic digestion can be recycled via heat exchangers to preheat the digester feed.

2.2.2.3 Temperature-Phased Anaerobic Digestion

Temperature-phased anaerobic digestion (TPAD) generally consists of at least one thermophilic stage followed by a mesophilic stage to improve digestion performance. Compared with conventional mesophilic digestion, TPAD generally provides better VSR, pathogen destruction, and gas production when treating combined solids (primary solids and WAS) (Han & Dague, 1996).

2.2.3 Anaerobic Digestion Pretreatment Processes

The goal of pretreatment is to improve digester performance, VSR, and pathogen destruction. Digestion pretreatment technologies can be classified as physical (e.g., ultrasound, pressure), thermal, chemical, and biological, or combinations thereof. Ideally, anaerobic digestion pretreatment should: (1) preserve organic material in the biomass; (2) promote hydrolysis of bacterial cells; (3) avoid the formation of any toxic and/or inhibitory compounds, and; (4) be environmentally friendly and economically feasible (Choi & Kwon, 2019).

Digestion pretreatment processes are most commonly applied to the WAS in digester feed. Currently, a few patented technologies have been applied at full scale: CambiTHP™, Biothelys® (thermal pretreatment), Aspal SLUDGE™, Praxair® Lyso™ (microwave pretreatment), Biosonator (ultrasonic pretreatment), MicroSludge™, Lysotherm ™, and Cellruptor (high-pressure homogenizer pretreatment) (Panigrahi & Dubey, 2019). If WAS comprises all or most of the digester feedstock, it may be worth exploring WAS pretreatment processes via additional research and perhaps pilot testing. The most popular digestion pretreatment process currently is thermal pretreatment, including thermal hydrolysis and thermo-chemical hydrolysis as described briefly below.

Thermal pretreatment improves the hydrolysis rate and disintegration of substrates at high temperature and pressure by addition of steam and subsequent decompression. The refractory compounds, macro-molecules in the sludges to be digested are hydrolyzed into micro-molecules via thermal pretreatment, which improves dewaterability by releasing the water bound in the bacterial cells, disinfects by sterilization, and reduces exogenous pollutants (Panigrahi & Dubey, 2019). The main design and process parameters of thermal pretreatment include temperature, pressure, duration, and heat transfer.

The thermal hydrolysis process is designed by heating the thickened or dewatered sludge under high pressure for a short reaction period (150 to 220 °C, 1380 kPa, 20 to 30 minutes). Typically, it includes preheating sludge through a heat exchanger before entering the reaction tank, where steam is added to maintain temperature and increase pressure. Upon the completion of the batch reaction, the solids are depressurized and pumped through heat exchangers to lower their temperature before mesophilic digestion. Thermal hydrolysis has become a popular and cost-effective pretreatment process that achieves pathogen-free biosolids, enabling wider use of the biosolids product.

2.2.4 Co-Digestion of External Feedstock

The potential to maximize digester gas production is often limited by an imbalance in the carbon/nitrogen (C/N) ratio of the feed solids, and ammonia accumulation in digestion. Anaerobic co-digestion, that is, simultaneous digestion of two or more substrates (such as biosolids and food waste), is a feasible and cost-effective approach to resolve these drawbacks. The abundant organic wastes from municipal, industrial, and agricultural field are ideal co-substrates because they can help balance the substrate nutrient by adding carbon, which is typically beneficial to optimize the C/N ratio of anaerobic digestion in WRRFs. They can also help adjust pH and dilute potentially toxic materials to mitigate inhibition to methanogens, consequently improving the CH_4 yield. (Forbes et al., 2012)

For example, the municipally available organic waste fat, oil, and grease (FOG) byproduct of food establishments and home cooking is used as a co-substrate with sludge at some WRRFs to increase CH_4 production, because FOG typically provides better CH_4 yield potential than other organic wastes because of its high concentrations of carbon and VS that are readily convertible to VFAs during digestion. (Kabouris, 2011).

Quite a few WRRFs now operate co-digestion programs, mostly in Europe and North America. The economic benefits of offsite organic wastes are gained through tipping fees and increased gas production, which can be substantial. Water resource recovery facilities practicing co-digestion have reported increased digester-gas production ranging from 10% to more than 75% after adding offsite feedstock, with some WRRFs generating almost all the energy needed to treat the wastewater flow to the WRRF.

Consideration must be given to many details in the design of offsite waste receiving, storage, heating, mixing, and digester-feed equipment. Improvements to the digestion facilities themselves must also be considered to facilitate better mixing and digestion of the feedstock, as well as improvements to digester-gas collection, storage, and treatment systems. These co-digestion system considerations are covered in other references (WEF et al., 2012; WEF et al., 2018). Digester gas system details are addressed in succeeding sections of this chapter.

3.0 DIGESTER GAS CHARACTERISTICS

This section describes the characteristics of digester gas produced by the anaerobic digestion process.

3.1 Primary Constituents of Digester Gas

As described in the previous section, the major constituents in digester gas are CH_4, CO_2, and water vapor ranging from 5% by volume at 35 to 37 °C (95 to 98 °F) to 15% by volume at 54 to 57 °C (130 to 135 °F). Small amounts of nitrogen and oxygen are also present because of air intrusion into the digester. Water vapor in digester gas will condense at ambient temperatures and can be further reduced with chilling. The temperature of digester gas must be considered to evaluate the composition of digester gas for end use.

3.2 Potentially Harmful Contaminants

Contaminants may be present in digester gas that may be volatilized from the liquid phase to the gas phase from a combination of physical and biological factors. Contaminants of concern are discussed in this section. The most common contaminants that require treatment are H_2S and siloxane compounds. Ammonia and other contaminants are also noted below.

3.2.1 Hydrogen Sulfide

Hydrogen sulfide occurs in digester gas from the reduction and volatilization of sulfur compounds present in the digester feed. It is a corrosive and harmful gas because it readily forms sulfuric acid in the presence of water. Typical concentrations of H_2S in digester gas range from less than 100 parts per million by volume (ppmv) to 2000 ppmv, although H_2S concentrations up to 12 000 ppmv have sometimes been observed in digester gas. The lowest end of this range may not be harmful, but the middle and higher ends of the range pose substantial threats to digester-gas handling and utilization because of severe corrosion of equipment created by acidic conditions. High H_2S concentrations in confined spaces also constitute a serious and sometimes lethal health hazard to WRRF operation and maintenance (O&M) personnel.

3.2.2 Siloxanes

Siloxanes are a class of organic silicon compounds composed of carbon, hydrogen, oxygen, and silicon that enter wastewater from domestic and industrial sources. Siloxanes are present in cosmetics and personal care products and may be discharged from silicon polymer manufacturing. Some siloxanes volatize in the anaerobic digester and become constituents of digester gas. The most common species of siloxane in digester gas are as follows (scientific name followed by common trade name in parentheses): octamethyl-trisiloxane (L-3), octamethyl-cyclotetrasiloxane (D-4), and decamethyl-cyclopentasiloxane (D-5). Siloxanes create problems with equipment that uses digester gas for fuel, because upon combustion, the siloxanes form silica deposits on metal engine components that are extremely difficult to remove.

3.2.3 Ammonia

Ammonia formed from the breakdown of proteins remains primarily in the aqueous phase at the typical pH ranging from 7.2 to 7.8 in most anaerobic digesters. The additional alkalinity created by ammonia typically helps to stabilize the anaerobic digestion process, but high concentrations of ammonia in the digesting sludge may also inhibit the digester process and release ammonia into the digester gas, which should be removed before use.

Formation of ammonia in digestion also has important implications to energy requirements at a WRRF. When centrate or filtrate from biosolids dewatering after digestion is returned to the head of the WRRF for retreatment, the ammonia content of the recycled return stream may create a significant demand for oxygen in the biological treatment process. This additional oxygen demand must be considered while evaluating digestion alternatives for a WRRF. Sidestream treatment ammonia removal or alternative mainstream treatment processes may need to be considered to reduce ammonia concentrations from the dewatering return stream.

3.2.4 Other Contaminants

Digester gas and biogas from other sources such as landfills may contain particulates and volatile organic carbon compounds (VOCs) that may need to be removed for beneficial use of the gas as a fuel. Particulates can be removed with sediment traps or in-line filtration that is typically the first step in a gas treatment system. Volatile organic carbon compounds are not typically found in digester gas at concentrations high enough to be of concern, but they can reach high concentrations in landfill gas and they tend to damage components in gas treatment and utilization systems. Volatile organic carbon compounds can also contribute to air pollution if not removed by scrubbing or by gas combustion.

4.0 DIGESTER GAS HANDLING AND SAFETY

This section provides information on safety considerations, including gas safety equipment, waste gas burners, safety regulations, and conveyance considerations including digester gas storage.

4.1 Introduction

Digester gas creates several challenges with its conversion to energy because of some of its constituents and variations in production rates. Most of the challenges are related to potential corrosive properties of digester gas, but because digester gas production and composition are variable, maintaining consistent flow and combustion rates may be challenging as well.

Digester gas from a healthy anaerobic digestion process generally consists of approximately 65% to 70% CH_4 and 30% to 3% CO_2 on a dry volume basis with trace amounts of nitrogen, hydrogen, H_2S, and water vapor. Formation of sulfuric acid from the reaction of H_2S and water vapor is largely responsible for corrosion of digester gas-handling equipment.

Low heating value (LHV) is the heat (energy) given off during combustion when the products of combustion remain in the gaseous state. In a bomb calorimeter, the products of combustion are cooled to the temperature of the original gas and air mixture. When this cooling takes place, the water produced during combustion condenses and the heat of condensation is added to the heat of combustion (LHV). The result is the high heating value (HHV). This heat of condensation can be calculated because the mass of water produced during combustion is known. The LHV should be used in the design and sizing of digester-gas management systems.

Digester gas typically has a LHV of 20 500 kilo-joules per cubic meter (kJ/m^3) to 24 200 kJ/m^3 (550 to 650 British thermal units per cubic foot [Btu/cu ft]) at 1 at 14.7 lb per square inch gravity (psig) and 15.5 °C (60 °F). It is an appropriate boiler fuel when burned with the proper type of burner and boiler. With an octane rating of approximately 120, digester gas is a suitable fuel for engines with high compression ratios and relatively high efficiencies, also known as "lean-burn" engines.

4.2 Safety Considerations

Safety considerations should be considered when designing digester gas management systems. The safety equipment serves multiple purposes, among which are to control pressure, prevent against explosion, flare excess gas, and/or protect the gas collection and storage systems from over- or under-pressurization.

4.2.1 Over/Underpressure

The containment of digester gas must be designed to prevent mixing of gas and air. This is done by maintaining positive gas pressure at all points in the digester-gas management system. Compressor suction piping must also be operated above atmospheric pressure. Alarms and/or shutdown features should be incorporated where any gas compressor may pump the system down to a vacuum or where changing liquid levels may create a vacuum. Most digester gas systems are designed to operate such that a pressure relief valve will vent gas and send an alarm at the highest system operating pressure. All digester roofs must be fitted with pressure and vacuum relief valves to limit the structural loading that could result from overpressure or vacuum.

Sediment and moisture (or drip) traps remove solids and moisture from the gas flow. These traps must be inspected and cleaned on a regular basis to ensure proper operation. It is recommended to locate sediment and moisture traps at all low points in digester-gas pipelines. Blockages and restrictions in gas collection systems are commonly caused by insufficient moisture and sediment traps, resulting in failure to remove sufficient condensate or grit from the gas piping. Moisture and sediment traps can be manual, electrically actuated, or continuous. Manual moisture traps are more commonly used, especially in indoor locations, partly because automatic moisture traps require frequent maintenance. It is recommended to install all moisture and sediment traps indoors, especially in regions experiencing cold winter seasons.

Low-pressure check valves should be installed only on the suction side of compressors to prevent flow reversal of the digester gas, causing a safety concern. As with all valves, they should be specified according to their performance requirements and capacities.

4.2.2 Flame Protection

Flame trap assemblies are used to stop the gas flow and snuff out a flame. The flame trap assembly consists of a flame arrestor on the upstream side and a thermal shutoff valve on the downstream side. The flame arrestor absorbs heat from a flame to drop the temperature below ignition. They are always installed upstream of a boiler or engine gas train and upstream of a waste gas burner.

Flame traps prevent the propagation of flame within the digester gas piping. These should be located close to the waste gas burner and downstream of the anaerobic digester. Sizing of the flame trap is based on maximum gas flowrate and allowable pressure drop.

4.2.3 Waste Gas Burners

Waste gas burners, also known as flares, are safety devices used to combust excess digester gas. Waste gas burners can be classified as conventional or enclosed combustion. Conventional flares, commonly known as candlestick or open flares, consist of a gas burner head, gas supply piping, and gas piping safety accessories, and an all-weather gas pilot and ignition system to ignite the mixture of waste gas and air. Combustion controls should include alarms for pilot failure and automatic pilot reignition system. Candlestick flares are open waste gas burners, exposed to atmospheric disturbances such as wind and precipitation and, as such, may have relatively poor performance.

Enclosed combustion flares provide an air-to-fuel ratio to achieve more complete gas combustion. They are typically installed in areas where flame stack emissions are highly regulated or where it is desired to hide the flame. Enclosed flares include more sophisticated controls to achieve complete combustion. However, they have a much higher capital cost than open flares.

4.3 Safety Regulations

The following North American Standards are recognized as either best design practice or, in some jurisdictions, code. Though not comprehensive, these documents should be referenced and adhered to when designing digester gas systems.

4.3.1 ANSI/CSA149.6

This document is a Canadian Standard originally developed in the late 1970s, with revisions every 5 years. In 2015, this was adopted by Amerian National Standards Institute (ANSI) as a joint standard between the United States and Canada. This document primarily covers equipment, materials, safety, and installation requirements.

4.3.2 NFPA 820

NFPA is the National Fire Protection Association. This document primarily covers area classifications for electrical equipment, ventilation of buildings, and provides guidance on equipment distances and use of flammable materials.

Generally, the only building areas that are designed with a hazardous classification for electrical and ventilation purposes are those where a combustible gas and air mixture is expected to occur, either occasionally or routinely. Explosion-proof design principles apply in rooms where digester gas mixing compressors and manifolds are located because it is expected that filter canisters, piping, and compressors will be periodically opened for maintenance.

The minimum ventilation in these areas for heat relief and code design typically exceeds the requirements for ventilation to declassify the space as hazardous. Separation of classified spaces from ordinary rooms will always require some fire-rated wall without doors and total closure with the common floor and ceiling. Pipe penetrations must be gas-tight.

4.3.3 Other Regulations and Guidance

Water Environment Federation Manual of Practice No. 8 (WEF et al., 2018) contains detailed information and guidance on digester-gas safety and equipment. Federal regulations such as 40 CFR 60.18 describe limits for air emissions and requirements for flares. Local regulatory agencies must also be consulted regarding air emissions and flares.

4.4 Piping Considerations

The first step during the design of the gas piping is to determine the pipe sizes based on the digester gas production rate at the digester, as described in Section 2.0. The digester gas production rate is typically determined from the solids loading rate, the volatile suspended solids (VSS) to total suspended solids (TSS) fraction, the percent volatiles destroyed, and the yield of digester gas to VSS destroyed. Typical digester-gas yields and piping design criteria are presented in Table 10.2.

Digester gas production rates fluctuate throughout the day, so digester-gas piping must be sized to accommodate for fluctuations in gas production rate. Depending upon operational parameters, digester-gas piping may need to be sized to handle flows 2 to 3 times the anticipated hourly gas production rate. After the maximum flow has been determined, the pipe size can be determined based on velocity. The maximum velocity in sludge-gas piping is typically 3.66 m/s (12 ft/sec). Staying below this velocity will reduce the carry-over of moisture from condensate traps and avoid high pressure losses and damage to other appurtenances.

It is not recommended to size digester gas piping and handling systems based on natural gas. Natural gas has a heating value higher than digester gas and, therefore, requires less gas to fuel a system. Because gas-pressure losses are based on the square of the gas flowrate, and because of its higher specific gravity, an equivalent fuel value of digester gas produces 2 to 3 times the gas-pressure loss of natural gas in a pipeline.

5.0 DIGESTER GAS CONDITIONING

The intended use of the digester gas dictates the required conditioning. To use digester gas as a vehicle fuel, for example, it is essential to precondition

TABLE 10.2. Digester gas production values and pipe design criteria

Parameter	Value
Digester gas yield	1.5–1.1 m^3/kg VSS destroyed (13–18 cu ft/lb VSS destroyed)
Maximum pipe velocity	3.66 m/s (12 ft/sec)
Minimum pipe slope	1.04 cm/m (0.125 in./ft)

the digester gas by removing water vapor, H_2S, CO_2, and other contaminants, leaving CH_4 at 97% or higher concentration.

Digester gas conditioning is a multistep process to remove constituents that are harmful to the equipment, the instrumentation, the conveyance system, and/or the environment. Each conditioning step is intended to alter the physico-chemical properties of the digester gas to make it amenable to processing in the next step with the ultimate goal of rendering the final quality of the digester gas suitable for the intended beneficial use. The quality of the digester feed sludge, the digester feed scheme, and the operational parameters of the digester all affect the quality and the production rate of digester gas. For example, the presence of siloxanes requires siloxane removal to protect downstream equipment such as intake valves, combustion engine heads, and membranes. A conditioning process might require a different conditioning process to precede it. For example, an activated carbon-based siloxane removal system requires that most of the moisture and H_2S be removed upstream, to ensure a longer service life for the carbon media.

The intended utilization of digester gas to recover energy, and the equipment used for energy recovery dictate the gas conditioning steps needed. The decision to condition digester gas to a desired quality should consider the capital, operation, and maintenance costs as well as permit conditions and environmental benefits. For example, digester gas from which H_2S, moisture, and siloxanes have been largely removed, and containing at least 50% CH_4 and up to 50% CO_2, is adequate for utilization in tractors, thus offering lower capital expense for small-volume digester gas systems (Owczuk et al., 2019).

5.1 Water Removal

Digester gas is saturated with water vapor at the time it leaves the digester. The concentration of water vapor in digester gas is initially in the range of 3% to 7% (13 to 83 g water per m^3 of digester gas). The presence of water vapor reduces the calorific value of digester gas, and it can cause damage to equipment downstream, such as compressors, boilers, engines, and carbon scrubbers.

When ambient temperature drops, digester gas cools, causing water vapor to condense. Condensate combines with CO_2 and H_2S to form acidic compounds causing accelerated corrosion and deposition of impurities on machines, gas scrubbers, pipelines, buffer vessels, sensors, and instruments. The removal of the water vapor from digester gas typically involves either absorption or adsorption. Absorption occurs when a dehydrating agent takes out the water vapor; whereas adsorption occurs when the water vapor is condensed and collected on the surface.

Refrigeration and absorption by a desiccant of moisture in digester gas are used to dehydrate the gas. In a refrigeration system, either an external chilled fluid system or an integrated air chiller can provide the cooling. The moisture condenses out of the digester gas as gas temperature reaches the dew point; the condensate trap releases the moisture into a drain. The incoming digester gas can be precooled using a gas-to-gas heat exchanger in which the treated gas is reheated by exchanging heat with the incoming digester gas. Cooling the digester gas from 38 °C (100 °F), the typical temperature of gas from an anaerobic digester, to below 5 °C (41 °F) can remove up to 75% of the water, as well as some impurities.

A desiccant dehydration system uses either a solid or a liquid desiccant dehydrator. The liquid desiccant dehydrator uses glycol, which has a chemical affinity for water, to absorb water vapor from the gas stream. The solid-desiccant dehydration consists of two or more adsorption towers filled with a solid desiccant. As the wet gas passes around the particles of desiccant material, water is retained on the surface of these desiccant particles.

5.2 Ammonia Removal

Digester gas may contain ammonia at varying concentrations. In combination with water, ammonia is highly corrosive and can damage pipelines. Because of ammonia's detrimental effect on digester gas quality, regulations on fuel purity demand precise reports on ammonia content before a gas is injected into the gas grid. A high concentration of ammonia in gas combustion engines leads to the formation of nitrogen oxides (NO_x, a regulated pollutant) in the engine exhaust.

Anaerobic digestion mineralizes nitrogen to ammonia and nucleic acids in the sludge biomass. The ammonia concentrations typically correlate directly to SRT and VSR; the concentration of ammonia increases, and organic nitrogen concentration decreases with longer SRT. Anaerobic digestion systems that use co-digestion of high fat and protein rich substrates to achieve higher quality gas and nutrient dense biosolids, stimulate the production of high ammonia levels, which can cause instability of the digestion process. Thermophilic anaerobic digestion at approximately 55 °C (131 °F) also results in higher ammonia concentrations in the anaerobic digester, and the ammonia becomes a significant component of digester gas.

A typical ammonia concentration in an anaerobic digester sludge is 0.1% by mass. Digester pH affects the concentrations at which ammonia becomes toxic, but ammonia concentrations reaching 0.3% by mass in the digester typically begin to inhibit digestion and VSR. Typical ammonia concentrations in digester gas are less than 10 ppmv. Systems using thermal hydrolysis typically produce ammonia concentrations an order of magnitude higher in the digester gas, in the range of 100 ppmv.

Low levels of ammonia in digester gas can be co-removed by dehumidification coupled with wet processes such as wet absorption process units. Higher levels of ammonia can be removed using acid spray scrubbers, biofilters, acidified clays, and addition of sodium hydrogen sulfate.

5.3 Hydrogen Sulfide Removal

Anaerobic microbes that break down organic matter in digestion produce trace gases including H_2S. The concentration of H_2S in digester gas ranges from less than one hundred to several thousand ppmv. The removal of gaseous H_2S reduces the potential for corrosion and human health hazards. The combustion of H_2S produces sulfur oxides (SO_x, another regulated air pollutant), which are major contributors to acid rain and air pollution. The technologies to remove H_2S vary in their level of treatment, capital, footprint, and O&M needs.

5.3.1 Physical-Chemical Systems

Physical-chemical treatment systems rely on media that react with the H_2S by converting/binding it. After time, when the media has reduced capacity to capture H_2S, it must be either regenerated, sometimes in situ, or removed/discarded and replaced with new media. Iron sponge and activated carbon are two types of physical/chemical treatment.

In an iron sponge process, digester gas is passed through a permeable bed of iron sponge (hydrated ferric oxide). For each mass unit of ferric oxide (Fe_2O_3) present in the system, 0.56 mass units of sulfide can be removed from digester gas. The hydrated media used is typically in the form of iron-dipped wood chips, soaked in water. As H_2S passes through the iron media, H_2S is converted to ferric sulfide and water in a passive chemical reaction. Alkaline conditions must be present to support this reaction, with a pH greater than 7.5. If the pH drops below 7, sodium bicarbonate is typically added to increase pH. The exhausted media is not hazardous and can be taken to any landfill for disposal. Iron oxide can be regenerated by adding air or oxygen; however, the regeneration is not complete and typically 30% of the reaction capacity is lost during each regeneration cycle.

A practical H_2S removal system consists of two stainless steel vessels piped to operate in a lead-lag configuration. Each vessel is sized to treat the full system flow rate of digester gas. The vessels are designed to operate in series, although the vessels are typically piped so that they can also be operated in parallel. During normal series operation, the lead vessel provides the bulk of the treatment and the lag vessel operates as a polishing step. When H_2S breakthrough is detected, the lead vessel is removed from service for media replacement. After media replacement is completed, the vessel with new media is placed in the lag position. An H_2S removal system with regenerative media can be regenerated somewhat while it is removing sulfur by adding oxygen to the feed gas stream.

5.3.2 Biological Systems

Biological treatment uses biological populations to convert H_2S to sulfuric acid and elemental sulfur that can be removed in a liquid and/or solid form. These systems use reactors filled with synthetic media conducive to biological growth, and the media requires periodic cleaning. Water, air, and nutrients are added to support biological growth. A general rule is that if gas inlet concentrations of H_2S are greater than 1000 ppm, then biological treatment systems become economically feasible. Typical outlet concentrations are 100 ppm or lower. Although the biological process is not new, its application to digester gas systems is relatively new, yet growing rapidly. In contrast to the adsorption type systems above, the biological systems are sensitive to varying influent loads and temperature.

5.3.3 Chemical-Biological Systems

The chemical-biological treatment uses at least two reaction vessels in series, the first being an absorption reactor and the second being a biological reactor. The absorption reactor facilitates the conversion of gaseous H_2S to aqueous sulfide, $H_2S(aq)$, through the addition of water and caustic chemical, used to raise the pH. The bioreactor then uses biological populations to oxidize dissolved sulfide, like biological treatment units described above. Chemical-biological systems can achieve efficient treatment for high H_2S concentrations, (e.g., 3500 ppm). However, compared to the biological system, it requires more vessels and is more operation intensive.

Biotrickling filters are irrigated reactors with a packed bed colonized by sulfide oxidizing bacteria, where a nutrient water "trickles" through the packed media from the top. As raw digester gas is blown through the media, H_2S is removed from the digester gas and metabolized by the bacteria. Sulfur-laden nutrient water is recirculated, and periodically discharged and replenished with makeup water. Air and recycled nutrient water sustain the biotrickling filter bacteria.

The trickling nutrient water must be delivered evenly to the media to prevent localized drying and channeling of digester gas. Spray nozzles must resist corrosion and fouling while delivering a mist small enough to provide large surface area, yet large enough to be captured by the demister or other water removal devices (e.g., chillers). Nutrient water is recirculated to reduce water usage. When the concentration of accumulated breakdown products (e.g., sulfates) becomes too high, a portion of the recirculating water must be replaced with fresh water. This can be controlled by a time setpoint, pH meters, and/or electrical conductivity meters.

5.3.4 Digestion Process Additives

Another way to remove H_2S from digester gas is to add a compound into the digester influent that will directly react with the H_2S produced before it has an opportunity to leave the digester. Such compounds are typically added to the digester feed solids before entering the digester. Overall, one advantage digester influent additives have over digester-gas scrubbing is lower reduced capital and maintenance and repair costs. Chemical options include liquid ferric chloride and powdered ferric hydroxide as described below.

5.3.4.1 Ferric Chloride Dosing

Ferric chloride can be injected directly into a digester by an automated dosing unit but is more likely added directly to the digester feed stream. The ferric chloride reacts with H_2S to form insoluble iron sulfides, which remain in the digested biosolids. Ferric chloride dosing is good for treating high initial H_2S concentrations as the first step in a multistage removal process. Initial costs are low because it does not require much equipment, but the operating cost can be high because of chemical cost, and the iron precipitates also add to biosolids mass and related handling costs.

5.3.4.2 Ferric Hydroxide Dosing

Ferric hydroxide, a dark orange compound, oxidizes reduced sulfurous compounds in the digester feed that helps reduce H_2S concentrations in the digester gas. Ferric hydroxide can remove H_2S from aqueous systems through oxidation and precipitation reactions within the operating pH range of most digesters (6.5 to 8.5). Ferric hydroxide typically requires a constant daily dose to consistently reduce the H_2S concentration. The required quantity of ferric hydroxide is based on its stoichiometric ratio to the volumetric flowrate of digester gas. Another advantage of using ferric hydroxide as an additive to the digester feed is the potential benefit of precipitating phosphorous into the digested biosolids. The reduced iron in the anaerobic environment binds with the dissolved phosphate, forming ferrous phosphate precipitate.

5.4 Siloxane Removal

Siloxanes in digester gas can be harmful to gas management systems in low concentrations; therefore, they are difficult to detect and control. Once formed, siloxanes are biologically stable, organic compounds. The damage caused by siloxanes and their combustion byproducts can be severely limiting to energy recovery, and it is costly to operations.

The manufacturers of energy recovery equipment and filtration membranes have generally set low limits for siloxanes in the parts per billion by volume (ppbv) range, but the limits can vary substantially depending on specific equipment. For example, microturbines are affected by siloxane concentrations as low as 5 ppbv, while some internal combustion engines can tolerate siloxane concentrations as high as 5000 ppbv (5 ppmv), so the end uses of digester gas dictate the extent of siloxane removal required.

In instances where H_2S concentrations are high, it is more economical to use an H_2S removal step ahead of the selective adsorption system for siloxanes. Siloxanes can deteriorate the performance of the costly gas filtration membranes if not removed upstream.

5.4.1 Carbon Based

Carbon-based removal systems have been the primary removal choice for siloxanes. Activated carbon is available that is manufactured for optimal siloxane adsorption. However, H_2S and VOCs in the digester gas compete with siloxane for adsorption sites. The carbon bed life can be increased by H_2S removal ahead of carbon adsorption. Volatile organic compounds are generally present in digester gas at low levels and are also removed by activated carbon, making activated carbon for digester gas treatment a single-use application, with the carbon being disposed of when exhausted.

5.4.2 Alumina Based

Alumina-based siloxane removal systems operate by passing the digester gas through a vessel containing activated alumina. This system consists of a twin vessel regenerating adsorption setup. The digester-gas stream is directed through one of the two vessels in which the siloxanes are adsorbed by the media, while the other vessel regenerates offline. Depending on the inlet concentration of siloxanes, each tank can be online for 24 to 72 hours. Regeneration of the media is achieved using a blower and electric heater; the collected siloxanes are exhausted to a thermal oxidizer. The regeneration takes approximately 12 hours. The switching of tanks typically occurs automatically on an adjustable cycle to provide continuous siloxane removal with no interruption in the gas stream.

5.4.3 Polymer Based

Polymer-based siloxane removal systems use specially formulated synthetic resins for capturing the siloxanes via surface adsorption. The polymeric adsorbents need to be regenerated.

5.4.4 Refrigerant Based

Removal of siloxane from the digester gas using refrigeration occurs by chilling and compressing the digester gas, which causes condensation of water vapor, and siloxanes adsorb onto the condensate. Chilling the digester gas to 5 °C (40 °F) and compressing to approximately 350 psi can remove 50% of the siloxane; further chilling the digester gas to −7 °C (−20 °F) can increase siloxane removal to 95%. (Wheless & Pierce, 2004).

5.5 Digester Gas Compression and Storage

The compression of digester gas is typically the central control point for effective transmission from production source in the digester to the intended end use. Compressor design must be coordinated with a wide range of factors in the digester gas system, including available suction pressure, desired discharge pressure, the amount and location of gas storage (if any), and discharge gas temperatures for downstream applications.

Because both the supply flow of digester gas and the demand at the end use can be constantly changing, the main purpose of the compressor system is to allow digester gas supply and demand to be matched. The compressor controls must first ensure that suction pressure does not fall below minimum limits (typically 2 to 3 in. of water column) to avoid pulling vacuum in the suction piping. The compressor controls must also ensure that discharge lines do not get overpressurized if the amount of gas supply exceeds what is being consumed from downstream processes. Pressure switches can be used to shut down compressors at low or high pressure setpoints, but this can lead to frequent on/off cycling causing maintenance concerns and problems for downstream processes requiring steady flow.

More sophisticated control systems include either variable-frequency drives (VFDs) ramping compressor speed in response to pressure setpoints on either the suction or discharge side of the compressor, or better yet in response to changes in gas storage levels on either the suction or discharge side of the compressor. Recirculation lines that use a three-way valve in the compressor discharge to return a portion of the discharge of the compressor to the suction can also be an effective way to control compression. These recirculation lines can typically react more rapidly and modulate more frequently than VFDs but also increase energy consumption and can elevate discharge temperatures. The most desirable compressor control strategy is a VFD and recirculation line used in combination to allow the compressor to ramp up and down in response to large swings in digester gas supply or demand, while the recirculation line is used for smaller and more frequent fluctuations in gas flow.

5.5.1 For In-Facility Uses Without Co-Generation

For facilities without co-generation, the most common use of digester gas is to combust it in boilers to create hot water or steam for digester and/or building heating. Required discharge pressures for digester gas feeding boilers will vary depending on distance from compressors to boilers and pressure regulating devices at the boiler but are typically on the order of 2 to 7 lb/sq in. gauge (psig). Because little to no gas conditioning is typically needed in conjunction with gas use in boilers, additional pressure to move gas through treatment vessels or heat exchangers are typically not needed.

Because digester gas production typically exceeds heating demands (particularly in the summer), the compressor controls can be simplified to include only provisions for providing a steady discharge pressure. As gas pressure builds up in the suction line because of on/off cycling, ramp down from a VFD, or discharge recirculation, then the flare on the suction side of the compressor can be used to control pressure.

The most common compressor types for the lower discharge pressure applications (2 to 10 psig) are *centrifugal* and *rotary lobe* compressors. Selection between these two compressor types should be based on suction inlet pressure requirements, discharge temperature range, compressor footprint, and maintenance requirements.

5.5.2 For Co-Generation Uses

For facilities using digester gas in co-generation with engines, required compressor discharge pressures are like the requirements of WRRFs without co-generation and are typically on the order of 2 to 10 psig. Additional pressure will be needed to move digester gas through treatment vessels and heat exchangers, but this additional pressure is typically on the order of 1 to 2 psig. For this reason, gas compression technologies used in engine co-generation systems are typically rotary lobe or centrifugal technologies as described in the previous section.

There are exceptions when higher discharge gas pressures are required for co-generation; these include instances in which the facility operates high pressure (~50 psig) gas storage spheres on the downstream or discharge portion of the compressors. Facilities that operate non-engine co-generation such as gas turbines or microturbines also require higher discharge pressures than reciprocating engines. The compressor options for these higher discharge pressure exceptions are discussed in the next section.

Because digester gas production typically significantly exceeds heating demands (particularly in the summer) the compressor controls can be simplified to include only provisions for providing a steady discharge pressure.

As gas pressure builds up in the suction line because of on/off cycling, ramp down from a VFD, or discharge recirculation, then the flare on the suction side of the compressor can be used to control pressure.

A schematic diagram of a typical digester-gas conditioning train for co-generation is shown in Figure 10.2.

5.5.3 For Vehicle and Pipeline Uses

For WRRFs that are upgrading digester gas to renewable natural gas (RNG) for vehicle and pipeline uses, as well as WRRFs using gas turbine or micro-turbine co-generation, higher compressor discharge pressures on the order of 100 to 200 psig are needed. Required discharge pressures will vary less on the distance from the compressor to the use point and will depend more on the upgrading or co-generation technologies being used. Microturbines typically require the lower range of pressure inputs, at approximately 75 to 100 psig. Pressure swing adsorption gas-upgrading technology typically requires approximately 100 to 125 psig. Membrane upgrading technology typically requires approximately 200 psig. Gas turbine co-generation typically requires 200 to 250 psig with high-pressure fuel storage between the

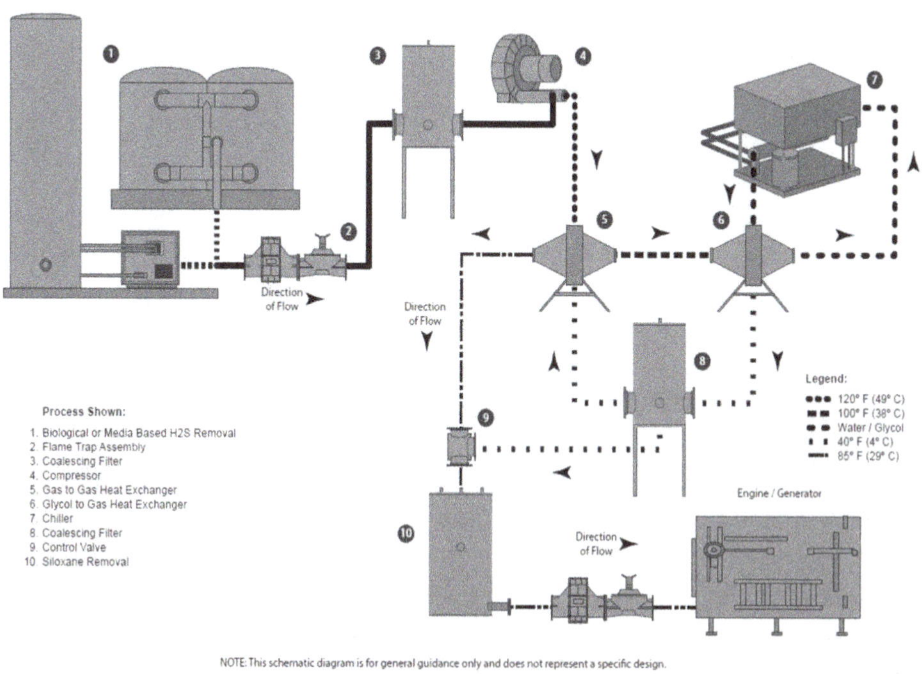

FIGURE 10.2. Typical digester-gas conditioning train for co-generation. Reprinted with permission from Varec Biogas, Inc.

compressor and prime mover. Because gas conditioning is needed in conjunction with microturbine and gas turbine co-generation, special design considerations need to be taken in gas conditioning design to allow for the treatment of high-pressure, compressed gas downstream of the compressor.

As noted in the previous sections, there are also additional applications that could require high-pressure gas compression on the order of 50 to 75 psig. These could include the use of high-pressure steel storage spheres, the use of diesel ignited co-generation engines, or simply transmitting digester gas over long distances. In these cases, the compressor types discussed below are also applicable.

Compressor types for the higher discharge pressure applications (50 to 200 psig) are (1) sliding vane, (2) liquid ring, (3) flooded screw, and (4) reciprocating compressors. Selection among these high-pressure compressors should be based on maximum discharge pressure, suction inlet pressure, consistency of output, discharge temperature range, power requirements, turndown capability, maintenance, and handling.

5.5.4 Digester Gas Storage

Digester gas can be stored at a wide range of pressure conditions, either pre- or post-compression. The volume of gas storage needed to satisfy a given energy need is inversely proportional to the storage pressure. The most common and reliable type of low-pressure digester gas storage at a WRRF is in the headspace of anaerobic digesters. There are two general types of onsite gas storage commonly used in a WRRF as described below.

5.5.4.1 Constant Volume, Variable Pressure Storage

This type of digester gas storage is essentially a medium- to high-pressure vessel that stores compressed gas. A pressure-regulating valve is installed downstream of the gas pipe that slowly releases gas at desirable operating pressure and flow to the end user. As gas stored in the tank depletes, the pressure in the storage vessel decreases and eventually stops gas releasing when pressure drops to a given setpoint (typically close to the operating pressure of the end user).

Gas is typically compressed to between 10 to 50 psig in WRRFs. Pressure vessels are made of steel or stainless steel and may theoretically be almost any shape, but shapes made of sections of spheres, cylinders, and cones are typically used. A sphere has the strongest structural integrity because of its spherical shape that offers uniform stress resistance, allowing the vessels to economically contain internal pressures. They require less land area yet provide more capacity than other pressure storage vessels, resulting in lower associated costs for piping, foundations, accessories, and painting. However, design and fabrication of this type of geometry is challenging and there have been patent disputes. An alternate design is a cylinder with end caps called *heads*. Head shapes are frequently either hemispherical or dished-shaped.

5.5.4.2 Constant Pressure, Variable Volume Storage

This storage type uses a mechanism that allows expansion of the tank volume when storage is required and contraction when gas depletion occurs, typically at low pressure matching the gas pressure in the digester of 10-in. to 15-in. water column. Storage solutions that apply this principle are: (a) gas holding steel cover on a concrete tank, (b) double membrane gas holder, and (c) gas bladder in steel or concrete tank, described as follows:

(a) *Gas Holding Steel Cover on a Concrete Tank:* This is an established technology, common to many anaerobic digestion systems. It is typically used to replace the digester roof in retrofit applications. The gas holder is a floating cover that could be of shell-theory dome or radial beam structure, with added side sheet and ballast to maximize stability. The extended skirt moves up and down vertically within the tank depending on gas volume. To increase stability during high winds, guides are installed to the cover sides, engaging guide devices in the digester wall. The guides are designed to permit either vertical or spiral movement of the cover.

(b) *Double Membrane Gas Storage:* This storage technology started in the farming industry in which low-cost biogas harvesting is a common practice. Because of its low cost, easy installation, and low maintenance, there are now several hundred units installed at WRRFs worldwide. It consists of an external membrane that forms the outer shape of the tank plus an internal membrane and a bottom membrane that make up the actual gas space. An air blower provides air to the space between inner and outer membrane, and thus keeps the gas pressure constant under all conditions. The pressurized air keeps the outer membrane in shape to withstand wind and snow loads, and it exerts a constant pressure on the inner membrane and thus pushes gas at constant volume and pressure into the outlet pipe. A safety valve is mounted on the inlet gas header to prevent the gas holder from overpressure. A pressure control valve is installed on the discharge header. Pressure in the holder is kept constant by allowing the gas holder to inflate or deflate. The storage is available as a semispherical dome on a concrete tank (typically replaces the roof of a digester) or as a standalone unit mounted on a concrete slab.

(c) *Gas Bladder in a Steel or Concrete Tank:* This storage essentially acts like a bladder tank, where the bladder itself is made of high-tensile strength polyester or closely woven nylon fabric, specially designed to be chemically resistant and flame resistant. The bladder is installed in a steel or concrete tank, mounted to an attachment bar around the tank shell and to a floating ring. It also includes a stabilizing weight to secure position. Gas withdrawal and gas filling are most often done via gas connections in the rigid bottom or top surface areas. The filling levels are measured by means of an ultrasonic level transmitter.

5.6 Upgrading Digester Gas to Fuel for Vehicles and Pipeline Injection

As noted previously, digester gas is composed primarily of CH_4 and CO_2. By separating and capturing CH_4 from digester gas, the resulting product is similar in composition and energy content to natural gas. Natural gas is a fossil fuel formed eons ago from degradation of organic matter, and it contains small fractions of heavier hydrocarbons such as ethane, propane, and butane, which are typically not present in digester gas. These heavier hydrocarbons may provide additional energy value to natural gas.

Pure CH_4 contains 1012 Btu per standard cubic foot (scf) while natural gas energy content can range from 940 to 1200 Btu per scf. To normalize these energy values, the fuel industry uses the Wobbe index to characterize the energy value and behavior of the fuel in a combustion device. The Wobbe index is defined as the energy content divided by the square root of the specific gravity of the gas compared to air. For example, compressed natural gas (CNG) vehicle fuel needs to have a Wobbe index between 1250 and 1400 (SAE J1616). For pipeline injection, the volumetric energy content may be specified with a minimum energy content between 950 to 990 Btu per scf. Digester gas upgrading systems can produce product CNG up to 99% CH_4 with CH_4 recovery from the digester gas also exceeding 99%. Upgraded digester gas may be referred to as CNG, RNG, or biomethane.

Pipeline injection standards are developed and promulgated by the owner of the natural gas distribution company. In most cases the local distribution company (LDC) controls access to the natural gas pipeline system. These companies are public utilities regulated at the state level. California, for example, has adopted statewide gas quality standards for injection of RNG into the natural gas system and has mandated that all gas utilities accept biomethane into their distribution systems. Alternatively, RNG can be injected into an interstate pipeline system that is regulated at the federal level. The gas quality standards for interstate pipelines may be less stringent than LDCs. In any case, the pipeline owner controls access and can shut down RNG injection at any time.

Compressed natural gas vehicle fuel typically has a lower energy content standard than pipeline injection. If the fueling system is owned and operated by the owner of the digester gas (typically the WRRF), there is no outside entity controlling the distribution of the product. However, wastewater utilities may find it difficult to find a vehicle fuel offtake for all their gas production, so pipeline injection is a more popular way to convey the RNG or CNG to its destination market.

Digester gas must be treated to remove moisture, contaminants, and CO_2 to be suitable for CNG vehicle fuel or pipeline injection. Generally, moisture and contaminants are removed by pre-treatment ahead of the CO_2 separation step, as described in preceding sections.

Three basic approaches have been used in the United States for digester-gas upgrading:

- pressure-swing adsorption,
- membrane separation, and
- water scrubbing.

There are many published studies comparing these technologies and for any application it may be possible to combine these technologies for an optimal system. In general, membrane separation lends itself to a modular design and is more cost-effective for small systems. Pressure-swing adsorption and water scrubbing benefit more from economy of scale. Each application needs to be evaluated taking into consideration site specific factors and the gas composition.

6.0 USES OF DIGESTER GAS FOR ENERGY RECOVERY

6.1 Introduction

This section provides an overview of combined heat and power (CHP) technologies. Options for upgrading digester gas into RNG for natural gas pipeline injection and compressed RNG for vehicle fuel are also discussed. This section includes a brief process description of each technology, typical performance characteristics, advantages and disadvantages, along with recent advancements, operational insights, and future trends.

A general schematic diagram of commonly used and potential digester-gas utilization options is shown in Figure 10.3.

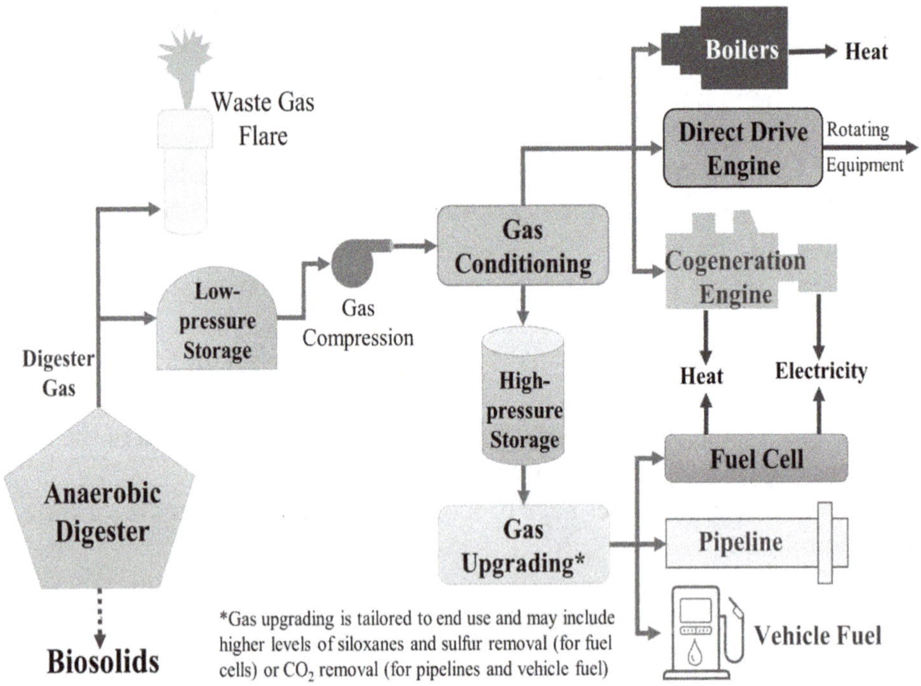

FIGURE 10.3. General schematic of digester gas utilization options. Reprinted with permission from Jacobs.

6.2 Heat from Digester Gas

Boilers extract usable heat energy from a fuel, typically via combustion. Historically, this is the most common technique for capturing energy from digester gas. The heat produced by boilers may be in the form of steam or hot water, depending on boiler type and temperature. Boilers designed to keep the anaerobic digestion process heated throughout the year are the most common use of digester gas. Boilers may also be used to provide hot water and building heat at other parts of the WRRF. Even treatment facilities that use engine generators or gas turbines need boilers as standby or supplemental heating equipment. Boiler sizes range from approximately 100 000 kilojoules per hour (kJ/h) to more than 1 million megajoules per hour (MJ/h).

Emissions controls are becoming increasingly important features for boilers. Air quality regulations must also be considered as part of all boiler designs. Modified boilers are sometimes required to meet nearly all air quality regulations. Following are brief descriptions of the general types of boilers using digester gas for fuel at WRRFs.

Fire-tube boiler: Packaged fire-tube boilers are the most common type of boilers used in WRRFs. They are available in sizes ranging from 2 to

30 MJ/h, equivalent to approximately 2 to 30 million British thermal units per hour (MMBtu/hr).

Fire-box boilers: Fire-box boilers are a special type of fire-tube boiler with an oversized combustion chamber. This chamber may help properly combust the relatively low caloric content of digester gas. Fire-box boilers range in size from approximately 2 to 10 MJ/h (2 to 10 MMBtu/hr).

Water-tube boilers: Water-tube boilers are like fire-box boilers, except that the combustion chamber is a horizontal insulated gas-tight compartment containing multiple, water-filled steel tubes that are heated by the combustion gases. They contain less water internally, so that water tube boilers often can warm up quicker and start up faster than fire-tube boilers. Those with flexible tubes are particularly resistant to thermal shock and less vulnerable to siloxane-caused silicon deposits on the tubes. Water-tube boilers are available in sizes ranging from less than 8 to 20 MJ/h (8 to 20 MMBtu/hr).

Cast-iron boilers: Cast-iron sectional boilers are sometimes used in retrofit applications because they can fit through doorways. These boilers are smaller, typically available in sizes from 300 000 kJ/h to 10 MJ/h (0.3 to 10 MMBtu/hr).

Digester gas can also provide some of the fuel needed for biosolids drying, which is discussed further in Section 6.6 below.

6.3 Direct Power from Digester Gas

Reciprocating engines, a type of internal combustion engine (ICE), are the most widely used technology in digester-gas direct power applications. With the goals of significantly better fuel economy and lower exhaust emissions, several manufacturers developed models that they used to create technically modern, progressive spark-ignition, lean-burn engines. With their higher fuel efficiency, these technologies enable WRRFs to replace or produce substantially more electric power to offset energy costs using existing digester-gas production.

At some WRRFs, reciprocating engines are connected directly to driveshafts for rotating equipment such as centrifugal blowers or pumps. These direct-drive engines save electrical power by replacing large electric motors, and the heat from the engine casings and exhaust systems may also be captured and used. The main disadvantage of using a direct-drive engine is that each engine's energy is dedicated to one equipment item, replacing one electric motor. If the treatment process at the WRRF does not require the driven equipment to operate, then the digester gas must be stored or flared. To overcome that disadvantage, most digester-gas fueled reciprocating engines generate both electrical power and heat, as described in the next section.

6.4 Co-Generation of Power and Heat from Digester Gas

Several types of systems can use digester gas a fuel for the co-generation of power and heat. The co-generation systems in use at WRRFs are described in this section (U.S. Environmental Protection Agency, 2006).

6.4.1 Internal Combustion Engines of Reciprocating Type

Internal combustion engines of the reciprocating type are the most prevalent technology used in CHP applications in WRRFs. Combining the energy converted to electrical power with recoverable heat from ICEs, also known as co-generation, typically represents approximately 60% to 70% of the inlet fuel energy. Jacket coolant and engine exhaust are the primary sources for this recoverable heat, which accounts for up to 30% of the recoverable fuel energy. A small amount of heat is sometimes recoverable from engine turbochargers and lube oil systems.

Jacket water cooling systems can produce hot water reaching up to 200 °F (93 °C) while engine exhaust temperatures range from 850 to 1200 °F (approximately 450 to 650 °C). To prevent corrosion of exhaust system components, engine exhaust outlet temperatures are maintained above the dew point (typically 250 to 350 °F [120 to 175 °C]) to prevent condensation. Engine exhaust heat available for recovery can generate hot water up to 80 °C (180 °F).

Typical characteristics of ICE-driven power generators fueled by digester gas are the following:

Electrical power output: From 200 kW to more than 5 megawatt (MW).

Thermal output: Can produce hot water or low-pressure steam up to 30 psig.

Fast startup capability: Allows ICEs to supply electricity quickly on demand.

Cold-start capability: As with other reciprocating engines, only batteries are required.

Availability: Typically, greater than 95% in stationary power generation applications.

Part-load operation: Ensures economical operation under variable electric loads.

Reliability and life: Proven as reliable for several decades given proper maintenance.

Exhaust emissions: Tend to be higher than emissions from other prime movers. The most significant pollutants found in ICE exhaust emissions are carbon monoxide (CO), oxides of nitrogen (NOx), and VOCs. When digester gas is used as fuel, oxides of sulfur and particulate matter (PM) are also present. Emissions from ICEs have improved in recent years through better design and control of the combustion process, and through use of exhaust catalysts.

Performance characteristics: The best digester-gas-fueled ICEs are based on lean-burn combustion; that is, the fuel-air mixture has considerably less fuel than its stoichiometric combustion ratio. By operating at lower air-to-fuel ratios, lean-burn engines operate with lower temperatures in the combustion chamber, which results in lower CO, NOx, and VOC emissions. Lean-burn engines also achieve higher fuel efficiency because of more complete fuel combustion, with reported energy-recovery efficiencies (including both electrical power and heat) of 70% to 85%. To meet most air-emissions permits, exhaust gas scrubbing systems are required.

6.4.2 Gas Turbines

Combustion gas turbines are another type of ICE. However, given the stark dissimilarities between combustion turbines and reciprocating engines, it is appropriate to define combustion gas turbines as a separate and distinct prime mover technology. Combustion gas turbines are often a good fit at large WRRFs (100 mgd or more).

Combustion gas turbines consist of three primary sections. The turbo-compressor compresses large quantities of atmospheric air. Fuel mixes with the compressed air within the combustion chamber and ignites. The combustion gases can reach temperatures of up to 1370 °C (2500 °F). The turbine, or expander, then extracts mechanical energy from the expanded, high-temperature gases, producing power and driving the turbo-compressor. Both the compressor and turbine sections consist of multiple stages of blades that rotate at high speeds.

Combustion gas turbines require less frequent maintenance than reciprocating internal combustion engines, but most WRRF staff are unfamiliar with this type of machinery, so maintenance of combustion gas turbines requires highly specialized service. When operated as a combined cycle, hot exhaust from the combustion gas turbine is directed to a heat recovery steam generator (HRSG), which produces the needed high-pressure steam that drives the steam turbine.

The main advantage of combined cycle systems is the ability to achieve higher electrical efficiencies, up to 50% or more. Combined cycle HRSGs produce superheated steam often exceeding 400 psig and 370 °C (700 °F), which typically requires full-time staffing of the HRSG by a certified onsite boiler operator. Usable heat that otherwise would have been available for digester heating or other uses is reduced because of the combined cycle's conversion of the turbine exhaust heat into electrical or mechanical power. For these reasons, combined cycle gas turbines are rarely used at WRRFs.

Heat energy is recoverable from combustion gas turbines equipped with an HRSG. In contrast to reciprocating internal combustion engines, which have multiple sources of potential heat recovery, heat recovery from combustion gas turbines is available only from the exhaust. Recovered combustion gas turbines heat can be used in the form of hot water, low-pressure steam, or high-pressure steam. Some of the newer combustion gas turbines include optional recuperators, which preheat combustion air. After passing through the recuperator, the turbine exhaust gases enter further treatment or the atmosphere.

Use of a recuperator enables a gas-turbine-based CHP system to operate at a higher electrical efficiency than without a recuperator. All combustion gas turbine systems report lower emissions with the use of a recuperator. The major drawbacks to recuperators are that they are more susceptible to damage caused by siloxanes in the digester gas. Therefore, digester gas fuel must be treated more thoroughly than for non-recuperated combustion gas turbines. The WRRF's power and heat generation objectives as well as cost, digester gas treatment, and emissions require careful evaluation when considering a recuperator as part of a gas-turbine based CHP system.

6.4.3 Steam Turbines

Steam turbines are among the oldest and most common of all large prime movers, especially for generating electricity. Steam turbines represent a distinct CHP technology that relies on thermal energy to produce mechanical and/or electrical power. Steam turbines are simple, reliable, and typically require little maintenance and steam boilers are also well proven.

Steam turbines are available as single-stage or multistage machines and in condensing or backpressure arrangements. In sizes below 10 MW, the steam turbine and boiler arrangement is too inefficient for cost-effective electric power generation. Often the greatest drawback to operating these systems at WRRFs is that boilers operating at steam pressures greater than 15 psig fall under the classification of power boilers and, therefore, require continuous, around-the-clock staffing by a licensed operator. This extra staffing requirement and the low operating efficiency of boiler and steam turbine CHP systems greatly limit their attractiveness in WRRFs.

Few WRRFs produce digester gas in adequate quantities to fuel this type of system, which has been installed at only one WRRF in the United States to date.

6.4.4 Microturbines

Microturbines represent a relatively new energy recovery technology especially suitable for smaller anaerobic digestion systems. They are small high-speed combustion gas turbines, most of which feature an internal heat exchanger like the recuperator on newer gas turbines described above. In a microturbine, a centrifugal compressor compresses the inlet air that is then preheated in the recuperator using heat from the turbine exhaust. The heated air from the recuperator mixes with fuel in the combustor, and hot combustion gas expands through the expansion and power turbines. The expansion turbine turns the compressor and, in single-shaft models, turns the generator as well. Two-shaft models use the compressor drive turbine's exhaust to power a second turbine that drives the generator. The recuperator uses the exhaust of the power turbine to preheat the air from the compressor.

Microturbines range in size from 30 to 200 kilowatts (kW). Microturbines are more complex than conventional gas turbines because of the recuperator, which introduces several additional internal pressure losses moderately reducing the system efficiency and power. The recuperator has four major internal connections, which pose a significant challenge to lower associated pressure losses while keeping manufacturing costs low and not compromising system reliability. Microturbines do not have high maintenance costs; however, they do have high fuel-treatment costs. Hence, their overall O&M costs can be proportionally higher because of their relatively small output. The electrical efficiency of microturbines typically ranges 26% to 27%.

6.4.5 Fuel Cells

Fuel cells are electrochemical devices that combine hydrogen with oxygen to continuously produce electricity. Fuel cells are the only CHP technology that does not rely on fuel combustion. Instead, fuel cells use electrochemical reactions to convert chemical energy to electricity and heat in a continuous, non-combustion manner. This is done by first extracting hydrogen from the fuel delivered to the unit while ambient air provides the necessary oxygen. The disassociation of the hydrogen and oxygen, with the formation of water, produces electricity.

The most notable fuel cell characteristics are high power generation efficiency and extremely clean exhaust emissions. Fuel cells operate without noticeable vibration and are quiet, although certain system accessories may cause minimal noise. In terms of the capital cost per kilowatt produced, fuel cells are the most expensive CHP technology. However, fuel cells have been particularly attractive in California because of that state's Self-Generation Incentive Program, which subsidizes renewable energy projects meeting the state's strict air-emission limits.

Four types of fuel cells exist: (1) phosphoric acid type fuel cells, (2) molten carbonate type fuel cells, (3) solid oxide type fuel cells, and (4) proton exchange membrane type fuel cells. However, only the molten carbonate type fuel cells (MCFCs) are currently available for use using digester gas as fuel. In contrast to the previously described CHP technologies that produce heat by fuel combustion, fuel cells are driven by electrochemical reactions.

The electrochemical reactions that occur within fuel cells are exothermic, meaning that heat is evolved from the reaction. In addition to heat, the electrochemical reactions produce water. The heat generated by the electrochemical reaction is enough to immediately vaporize the water formed during the same chemical reaction. Heat produced by the electrochemical reaction is available for digester heating.

Most fuel-cell-based CHP systems require a boiler to make up the shortfall of heat energy not provided by the fuel cells, thus maintaining the anaerobic digesters at the appropriate temperature. Boilers at WRRFs operate with either digester gas or natural gas. Therefore, the use of a boiler to supplement WRRF heating needs decreases the overall CHP system efficiency and does not maximize the energy available in digester gas. Molten carbonate type fuel cells operate at temperatures reaching approximately 677 °C (1250 °F). In addition to anaerobic digester heating, the recoverable heat from MCFCs is also capable of producing enough high-pressure steam to drive a steam turbine, generating additional power.

Fuel cells have the lowest exhaust emissions of all CHP systems, because the fuel conversion process does not include combustion. Instead, fuel cells rely on electrochemical reactions to convert fuel into hydrogen, while generating electricity and heat. Given their extremely clean emissions, fuel cell installations to date have not required air-emissions permitting.

6.5 Digester Gas Conversion to Renewable Natural Gas for Commercial Use

As discussed in the previous sections, many uses of digester gas for energy recovery at the WRRF are well-established. However, some WRRFs have preferred to use digester gas for commercial uses such as

- Producing RNG for injection into natural gas pipelines and
- Producing RNG for vehicle fuel.

A gas treatment (upgrading) system required to process the digester gas to RNG, as described in Section 5.6, must meet strict specifications before being accepted by the commercial market. Typical municipal digester gas composition compared with typical RNG specifications are shown in

Table 10.3. Analysis of digester gas to identify specific digester-gas composition with respect to specific commercial requirements for RNG will be necessary for design of digester-gas treatment systems for upgrading to RNG.

The treatment sequence to produce RNG from digester gas varies slightly between different manufacturers but generally includes gas pressurization, gas drying, H_2S removal, gas compression, CO_2 removal, thermal oxidization, and gas preparation.

Advantages of producing RNG from digester gas include accessing an alternative market for the digester gas, simplifying the digester and building heating systems, and reducing onsite emissions. In most of the world, there are also significant financial incentives for producing RNG offered in the way of carbon credits, such as the Renewable Identification Number (RIN) system set up by the U.S. Environmental Protection Agency in the United States. These regulatory-driven financial incentives often create favorable economics for investing in systems that upgrade digester gas to produce and sell RNG to commercial markets for vehicle fuels. Selling RNG to a natural gas company or other commercial fuel markets (distribution, wholesaler, etc.) allows the WRRF to participate in the supply of energy to the community.

Disadvantages of producing RNG from digester gas include increased capital and O&M costs for digester-gas upgrading equipment at the WRRF. In addition, producing and selling RNG eliminates a potential source of cost-effective and reliable electrical power for the WRRF. Issues associated with the marketability of RNG at WRRFs include fluctuating natural gas prices that determine the market value of RNG, and sometimes the

TABLE 10.3. Typical compositions of digester gas and RNG

Parameter	Digester Gas	Renewable Natural Gas (RNG)
Methane (CH_4), percent by volume	58–70	>97
Carbon dioxide (CO_2), percent by volume	25–39	<2
Oxygen, percent by volume	0–1	<0.1
Hydrogen sulfide (H_2S), ppmv	800–4000	4
Siloxanes, ppbv	Varies	20
Water, pounds per million standard cubic feet (scf) of gas	Saturated	7

reluctance of natural gas companies to accept RNG into their systems from untested or unproven providers.

6.5.1 Injection of Renewable Natural Gas into Natural Gas Pipelines

One commercial use for digester gas upgraded to RNG is injection into natural gas pipelines. For this use, digester gas must be thoroughly cleaned and pressurized before introduction into the natural gas supply. To achieve this, water, CO_2, and H_2S are removed from digester gas so that it approaches the purity of natural gas, as previously described in Section 5.6.

6.5.2 Renewable Natural Gas as Vehicle Fuel

Digester gas can also be upgraded to RNG and used as vehicle fuel, as previously described. For this use, digester gas is treated (including removal of most CO_2) and compressed for use in fleet vehicles or other equipment. For utilities that already use natural-gas-fueled vehicles, this alternative may be cost-effective. However, vehicle conversion, the construction of fueling stations, and digester-gas purification and compression equipment must be considered when evaluating this use.

6.6 Digester Gas as Heat Source for Biosolids Drying

Digester gas can also be used as fuel for a thermal biosolids dryer. Thermal drying will produce a biosolids product that typically contains more than 90% dry solids (no more than 10% water). This greatly reduces the potential pathogens, the total volume of biosolids, and the subsequent biosolids disposal and handling costs.

Digester gas would preferentially be used as fuel for boilers to heat the digesters, and any remaining gas could serve as a heat source for the thermal drying process. Without supplemental fuel, however, it is likely that not enough gas will be available to fully support both digestion and drying. This would be particularly true in the colder climates, when digester heating will require more fuel gas, making less available for drying.

Two types of dryers are commonly available:

- *Direct dryer*, in which the biosolids are heated directly by hot air in contact with biosolids, including rotary drum dryers, belt dryers, and fluidized bed dryers. Operating temperatures vary from belt dryers in the range of 200 to 300 °C to drum dryers in the range of 400 to 500 °C.

- *Indirect dryer,* in which the biosolids are heated by contact with hot surfaces such as paddles or screws flights, with operating temperatures in the range of 200 to 300 °C.

Digester gas combustion can be the heat source for all thermal dryers. With a direct dryer, air is heated directly from digester-gas combustion and is blown into the biosolids for drying. With an indirect dryer, a heating medium (such as steam or oil) is heated by digester-gas combustion. The hot steam or oil contacts and heats the surfaces that, in turn, heat and dry the biosolids.

The product of heat drying is dried biosolids that consist of approximately 8% to 10% water (90% to 92% solids) and handles like a granular fertilizer or blended soil. The use of digester gas for biosolids drying recovers the energy from digester gas, not only by replacing fossil fuel sources of heat, but also by reducing the energy needed for transporting and handling biosolids.

Other potential uses of digester gas for energy recovery may be addressed in Chapter 12—"Looking Forward."

7.0 REFERENCES

Banks, C. J., Zhang, Y., Jiang, Y., & Heaven, S. (2012.) Trace element requirements for stable food waste digestion at elevated ammonia concentrations. *Bioresource Technology*, *104*, 127–135.

Batstone, D. J., Keller, J., Angelidaki, I., Kalyuzhnyi, S., Pavlostathis, S., Rozzi, A., Sanders, W. T. M., Siegrist, H., & Vavilin, V. A. (2002). The IWA anaerobic digestion model no 1 (ADM1). *Water Science and Technology*, *45*, 65–73.

Bayr, S., Pakarinen, O., Korppoo, A., Liuksia, S., Väisänen, A., Kaparaju, P., & Rintala, J. (2012). Effect of additives on process stability of mesophilic anaerobic monodigestion of pig slaughterhouse waste. *Bioresource Technology*, *20*, 106–113.

Beall, S. S., Jenkins D., & Vidanage S. A. (1998). A systematic analytical artifact that significantly influences anaerobic digestion efficiency measurement. *Water Environment Research*, *70*(5), 1019–1024.

Choi, H. K., & Kwon, J. S. I. (2019). Multiscale modeling and model-based feedback control of pulp digester. In *Proceedings of the 2019 American Control Conference (ACC)*, Philadelphia, PA, July 10–12, 2019, pp. 2819–2824.

Forbes, R. H. (2013). *Review, update, and comparison of advanced anaerobic Digestion Processes*. Presented at the 90th Annual Conference and Exposition of the Water Environment Federation in Chicago, IL, October 2013.

Forbes, R. H., Kabouris, J., Jalla, S., & Gabel, D. (2012). *Key considerations for maximizing combined heat and power from anaerobic digestion by adding fats, oils, and grease and high-strength wastes*. Presented at the 89th Annual Conference and Exposition of the Water Environment Federation, October 2012.

Han, Y., & Dague, R. (1996). *Laboratory studies on the temperature-phased anaerobic digestion of mixtures of primary and waste activated sludge*. Presented at WEFTEC Conference, October 1996.

Kabouris, J. C. (2011). *I have my waste, now what do i do with it?—Process design*. Presented at the preconference workshop "Codigestion of High Strength Organic Wastes at Municipal Wastewater Treatment Plants" of the Water Environment Federation Residuals and Biosolids Conference, Sacramento, CA, May 22–25, 2011.

Owczuk, M., Matuszewska, A., Kruczynski, S., & Wojciech, K. (2019). Evaluation of Using Biogas to Supply the Dual Fuel Diesel Engine of an Agricultural Tractor. *Energies Journal-MDPI*, March 20, 2019.

Panigrahi, S., & Dubey, B. K. (2019). A critical review on operating parameters and strategies to improve the biogas yield from anaerobic digestion of organic fractionof municipal solid waste, *Renewable Energy 143*, 779e797. https://doi.org/10.1016/j.renene.2019.05.040

Shen, Y., Linville, J., Urgun-Demitras, M., Mintz, M., & Snyder, S. W. (2015). An overview of biogas production and utilization at full-scale wastewater treatment plants (WTTPs) in the United States: Challenges and opportunities towards energy-neutral wwtps. *Renewable and Sustainable Energy Reviews Journal, 50*. Journal ID: ISSN 1364-032.

U.S. Environmental Protection Agency, Combined Heat and Power Partnership. (2006). *Opportunities for and benefits of combined heat and power at wastewater treatment facilities*.

Water Environment Federation, American Society of Civil Engineers, & Environment & Water Resources Institute. (2018). *Design of municipal wastewater treatment plants* (6th ed.; WEF Manual of Practice No. 8; ASCE Manual and Report on Engineering Practice No. 76). McGraw-Hill.

Water Environment Federation, Water Environment Research Foundation, & U.S. Environmental Protection Agency. (2012). *Solids process design and management*.

Wheless, E., & Pierce, J. (2004). Siloxanes in landfill and digester gas update. *Proceedings of the 27th Annual Solid Waste Association of North America Landfill Gas Symposium*, March 2004.

8.0 SUGGESTED READINGS

Allison, S. D., & Martiny, J. B. (2008). Resistance, resilience, and redundancy in microbial communities. *Proceedings of the National Academy of Sciences, 105*, 11512–11519.

Amani, T., Nosrati, M., & Sreekrishnan, T. (2010). Anaerobic digestion from the viewpoint of microbiological, chemical, and operational aspects—A review. *Environmental Reviews, 18*, 255–278.

Bryant, M. (1979). Microbial methane production—Theoretical aspects. *Journal of Animal Science, 48*, 193–201.

Calli, B., Mertoglu, B., Inanc, B., & Yenigun, O. (2005). Effects of high free ammonia concentrations on the performances of anaerobic bioreactors. *Process Biochemistry, 40*, 1285–1292.

Capson-Tojo, G., Moscoviz, R., Astals, S., Robles, A., & Steyer, J. P. (2020). Unraveling the literature chaos around free ammonia inhibition in anaerobic digestion. *Renewable & Sustainable Energy Reviews, 117*.

Capson-Tojo, G., Ruiz, D., Rouez, M., Crest, M., Steyer. J.-P., Bernet, N., Delgenès, J-P., & Escudié, R. (2017). Accumulation of propionic acid during consecutive batch anaerobic digestion of commercial food waste. *Bioresource Technology, 245*, 724–733.

Chen, Y., Cheng, J. J., & Creamer, K. S. (2008). Inhibition of anaerobic digestion process: A review. *Bioresource Technology, 99*, 4044–4064.

Conrad, R. (1999). Contribution of hydrogen to methane production and control of hydrogen concentrations in methanogenic soils and sediments. *FEMS Microbiology Ecology, 28*, 193–202.

Demirel, B., & Scherer, P. (2008). The roles of acetotrophic and hydrogenotrophic methanogens during anaerobic conversion of biomass to methane: A review. *Reviews in Environmental Science and Bio/Technology, 7*, 173–190.

Forbes, R. H., De Andrade, B., & Fortin, A. (2018). *Exploring the feasibility of water resource and recovery facility (WRRF) biogas utilization in Brazil.* Presented at the 95th Annual Conference and Exposition of the Water Environment Federation in New Orleans, LA, October 2018.

Forbes, R. H., & T. Johnson (2011). *Calculating plant energy balance and carbon footprint using CHEApet (carbon heat energy assessment and plant evaluation tool)*. Presented at the Water Environment Federation's 2011 Energy and Water Specialty Conference in Chicago, IL. August 2011.

Forbes, R. H., Jalla, S., & Horton, J. (2010). *Fast track to new digester gas-to-energy facilities at the F. Wayne Hill Water Resources Center in Gwinnett County, GA*. Presented at the Water Environment Federation's 2010 Residuals and Biosolids Management Specialty Conference in Savannah, GA. May 2010.

Gonzalez-Martinez, A., Garcia-Ruiz, M. J., Rodriguez-Sanchez, A., Osorio, F., & Gonzalez-Lopez, J. (2016). Archaeal and bacterial community dynamics and bioprocess performance of a bench-scale two-stage anaerobic digester. *Applied Microbiology and Biotechnology, 100,* 6013–6033.

Kayhanian, M. (1999). Ammonia inhibition in high-solids biogasification: An overview and practical solutions. *Environmental Technology, 20,* 355–365.

Krause, L., Diaz, N. N., Edwards, R. A., Gartemann, K.-H., Krömeke, H., Neuweger, H., Pühler. A., Runte, K. J., Schlüter, A., Stoye, J., Szczepanowski, R., Tauch, A., & Goesmann, A. (2008). Taxonomic composition and gene content of a methane-producing microbial community isolated from a biogas reactor. *Journal of Biotechnology, 136*, 91–101.

Rajagopal, R., Massé, D. I., & Singh, G. (2013). A critical review on inhibition of anaerobic digestion process by excess ammonia. *Bioresource Technology, 143,* 632–641.

Raynal, J., Delgenes, J., & Moletta, R. (1998). Two-phase anaerobic digestion of solid wastes by a multiple liquefaction reactors process. *Bioresource Technology, 65*, 97–103.

Schink, B. (1997). Energetics of syntrophic cooperation in methanogenic degradation. *Microbioogy and Molecular Biology Reviews, 61*, 262–280.

Smith, A. M., Sharma, D., Lappin-Scott, H., Burton, S., & Huber, D. H. (2014). Microbial community structure of a pilot-scale thermophilic anaerobic digester treating poultry litter. *Applied Microbiology and Biotechnology, 98*, 2321–2334.

Tracy, B. P., Jones, S. W., Fast, A. G., Indurthi, D. C., & Papoutsakis, E. T. (2012). Clostridia: The importance of their exceptional substrate and metabolite diversity for biofuel and biorefinery applications. *Current Opinion in Biotechnology, 23*, 364–381.

Uludag-Demirer, S., Demirer, G. N., Frear, C., & Chen, S. (2008). Anaerobic digestion of dairy manure with enhanced ammonia removal. *Journal of Environmental Management, 86,* 193–200.

Vartoukian, S. R., Palmer, R. M., & Wade, W. G. (2007). The division "Synergistes". *Anaerobe, 13,* 99–106.

Walker, M., Iyer, K., Heaven, S., & Banks, C. (2011). Ammonia removal in anaerobic digestion by biogas stripping: An evaluation of process alternatives using a first order rate model based on experimental findings. *Chemical Engineering Journal, 178*, 138–145.

Wang, H., & Larson, R. (2015). *Effects of mixing duration on biogas production and methanogen distribution in dairy manure anaerobic digesters.* Waste to Worth: Spreading Science and Solutions. Seattle, Washington.

Wang, H., Fotidis, I. A., & Angelidaki, I. (2015). Ammonia effect on hydrogenotrophic methanogens and syntrophic acetate-oxidizing bacteria. *FEMS Microbiology Ecology, 91.*

Wang, P., Wang, H., Qiu, Y., Ren, L., & Jiang, B. (2018). Microbial characteristics in anaerobic digestion process of food waste for methane production–A review. *Bioresource Technology, 248*, 29–36.

Wang, X., Yang, G., Feng, Y., Ren, G., & Han, X. (2012). Optimizing feeding composition and carbon–nitrogen ratios for improved methane yield during anaerobic co-digestion of dairy, chicken manure and wheat straw. *Bioresource Technology, 120*, 78–83.

Water Environment Research Foundation. (2011). *State of the science on biogas: Treatment, co-generation, and utilization in high temperature fuel cells and as a vehicle fuel* (WERF Publication OWSO 10C10a).

Westerholm, M., Levén, L., & Schnürer, A. (2012). Bioaugmentation of syntrophic acetate-oxidizing culture in biogas reactors exposed to increasing levels of ammonia. *Applied and Environmental Microbiology, 78*, 7619–7625.

Ziganshin, A. M., Liebetrau, J., Pröter, J., & Kleinsteuber, S. (2013). Microbial community structure and dynamics during anaerobic digestion of various agricultural waste materials. *Applied Microbiology and Biotechnology, 97*, 5161–5174.

11

Renewable Energy

Hossain Azam, PhD, PE; Arifur Rahman, PhD, PE; Mahmudul Hasan; Lucas Botero, PE, BCEE, ENV SP; & Lirane Mandjoupa

1.0 INTRODUCTION

Water resource recovery facilities (WRRFs) are interested in implementing renewable sources of energy when and where it makes sense to implement it economically. Recovery of renewable energy at WRRFs together with other energy efficiency measures provides a powerful basis for achieving energy neutrality. Water resource recovery facilities can not only treat wastewater but also can recover energy from the wastewater itself; flow of wastewater and sludge (thermal/hydraulic); digester gas (CH_4, H_2); untapped spaces of treatment facilities for solar, wind, and thermal energy; and more from the vicinity of WRRFs. The potential of recovering solar, hydraulic, thermal, and wind energy at WRRFs can be effective, as it does not need large land areas as required for other types of renewable energy sources such as biomass/biofuel production from agricultural land. Properly treated wastewater can be used as a heat source in winter and can be used for cooling in summer. The energy can be used internally at the facility to become self-sufficient or to supply external infrastructures or the main electrical grid. Thus, WRRFs can be considered as a renewable energy supplier. Biogas, or anaerobic digester gas, a byproduct of anaerobic digestion, contains 60% to 65% methane and can be used to generate electricity and heat for on-site use, as described in Chapter 10.

Solar photovoltaic systems are being widely installed at WRRFs throughout the United States. Solar photovoltaic can be particularly attractive because it does not rely upon or interfere with the traditional WRRF's treatment processes. Wind turbines can be installed at WRRFs to generate power from wind, operating up to 24 hours per day in areas with naturally occurring consistent winds. Recovery of thermal energy and hydraulic energy are additional sources that may also be considered and have been implemented in WRRFs. Thus, renewable energy can be major energy support for WRRFs and can complement the energy need of WRRFs.

2.0 APPLICATION OF RENEWABLE ENERGY TECHNOLOGIES

2.1 Solar Energy (Photovoltaic)

2.1.1 Introduction

Many WRRFs are adopting solar energy to become energy neutral. Solar energy is a well-established renewable energy technology in the overall energy market in North America (Gikas, 2017; Maktabifard et al., 2018; Spellman, 2013a; Strazzabosco et al., 2019). Solar photovoltaic-generated electricity supplies baseline power needs at WRRFs throughout the United States. Solar photovoltaic takes advantage of large unused spaces such as building roofs, tank covers, or relatively flat, sun-exposed land in between structures or within buffer lands surround utility assets. For solar WRRFs without available area, an alternative is the use of remote photovoltaic panel arrays connected to the WRRF's energy grid (Gikas, 2017; Maktabifard et al., 2018; Spellman, 2013a; Strazzabosco et al., 2019).

A solar photovoltaic facility can generally be described as a collection of solar panels mounted on tilted racks. The solar panels are grouped together in series of modules called *strings* to achieve the desired direct current system voltage. These strings are then combined and fed to an inverter that converts the electricity from direct current to alternating current. The output of the inverter may be fed to a step-up transformer to match the interconnection voltage. The configuration and the equipment needed depend on the interconnection, type of inverter used (central or string), and if an energy storage system is used (Spellman, 2013a; Strazzabosco et al., 2019).

Solar photovoltaic installations generate power as long as there is sunlight, and the production is transient unless batteries are used to store energy when no light is directed to the panels. To produce 1 MW of electricity, a land-based system requires 2.02 to 2.43 hectares (5 to 6 acres) of land, and the land generally needs to be flat and clear of obstructions (Sunrun, n.d.). The different components of the photovoltaic system are shown in Figure 11.1. Some of the photovoltaic modules available in the market at the time of this publication are shown in Table 11.1.

Smaller scale solar energy utilization at WWRFs through direct solar equipment use, such as solar mixers, has also been implemented, as shown in Figure 11.2.

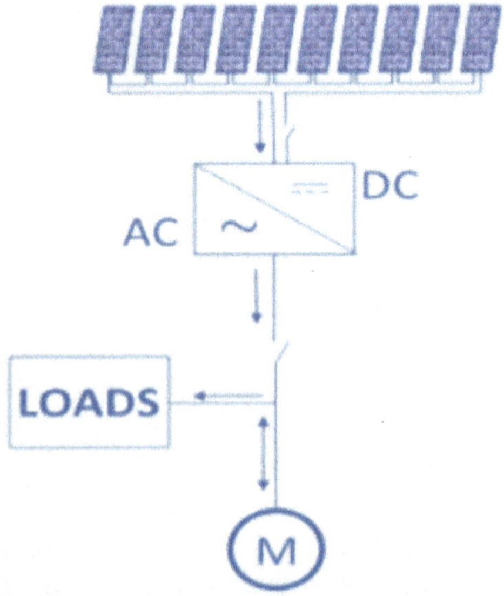

FIGURE 11.1. Photovoltaic system components. Reprinted with permission from Black & Veatch.

TABLE 11.1. Examples of different photovoltaic module types available in the market

Types	Notes
Monocrystalline	Higher capital cost High efficiency
Polycrystalline	Lower capital cost Average efficiency
CdTe thin film	Less efficient than polycrystalline
Bifacial	Power generation on both sides of module

2.1.2 Barriers and Incentives for Implementing Solar Energy at Water Resource Recovery Facilities

Water resource recovery facilities have used renewable energy generation (e.g., solar energy) to support the increasing energy demand associated with wastewater treatment, mitigating rising energy costs from the energy grid and associated greenhouse gas emissions (Strazzabosco et al., 2019). The following barriers and incentives have been identified to implement solar photovoltaic in WRRFs as shown in Table 11.2.

FIGURE 11.2. Solar mixer. Reprinted with permission from IXOM Watercare, Inc.

TABLE 11.2. Barriers and incentives of solar photovoltaic in WRRFs (Local Government Commission [LGC], n.d.; Solar Energy Industries, n.d.; Strazzabosco et al., 2019).

Barriers	Incentives
Nighttime energy demand coverage and high cost of energy storage systems	Tax incentives for solar power
Land availability	Public awareness towards renewable energy
Geographic location (site solar irradiance)	Economies of scale lower cost of photovoltaic panels
Low electricity rates	High utility rates
Lack of interest to reduce CO_2 emissions	Capitalize on CO_2 emissions reduction inceptives (tax or otherwise)
Environmental factors affecting solar panels (soiling)	Potential for high electricity rates in the future

2.1.3 Case Studies

Many WRRFs have adopted or in the process of adopting solar installations around the world. Few examples of such U.S. installations are shown in Table 11.3.

2.1.4 Cost–Benefit of Analysis of Solar Photovoltaic in Water Resource Recovery Facilities

The use of photovoltaic in WRRFs is highly site specific. Water resource recovery facilities that have implemented successful solar energy projects typically had one or more specific factors driving its implementation (LGC, n.d.; Sunrun, n.d.). Some of the important factors are:

- WRRFs with adequate land availability. Buffer areas can be used for solar photovoltaic deployment;
- Areas of high solar irradiance and flat terrain;
- High energy costs;
- Facilities with renewable goals in mind; and
- Facilities with high peak demands during the day that could benefit from load shaving using onsite generation through photovoltaic panels.

The economics of solar photovoltaic change from each location and energy market conditions. Costs for solar photovoltaic systems have decreased over the years, making this renewable energy option more attractive. The use of alternative ownership structures (e.g., land leases with power

TABLE 11.3. Examples of solar installations in different WRRFs in the United States

1) City of Las Vegas (CLV), Water Pollution Control Facility (WPCF), Nevada, USA	**WRRF Description**	The WPCF has been serving the CLV and City of North Vegas (CNLV) sewer-sheds since 1957 when the facility was first inaugurated. Expansions have been taking place over the years since the facility-rated capacity increased to 344 ML/d (91 mgd) annual average daily flow (AADF) in 2003. Average power demand for the facility is 5 MWh.			
	Drivers Photovoltaic for Generation	CLV wanted to take advantage of their renewable sources of energy and the land available at the WRRF to diversify energy inputs, minimize energy cost, and become more sustainable.			
	Photovoltaic Energy Generation Output	**Area of Panels**	**Types of Photovoltaic Panels**	**Types of Mounting**	**Types of Metering**
	3 MW alternating current (STC)	~13	Multicrystalline	Dynamic tracking	Smart meter, separate from the WRRF system
2) Southern Region WRRF, Palm Beach, Florida, USA	**WRRF Description**	The Southern Region WRRF is permitted to treat an AADF of 91.0 ML/d (24.0 mgd) of wastewater. Average power demand for the facility is 2500 KWh/mo.			
	Drivers Photovoltaic for Generation	The county wanted to take advantage of their renewable sources of energy and the land available at the Valrico WRRF to diversify energy inputs, minimize energy cost, and become more sustainable.			
	Photovoltaic Energy Generation Output	**Area of Panels**	**Types of Photovoltaic Panels**	**Types of Mounting**	**Types of Metering**
	162 KW alternating current (STC) proposed	12 Ac	Multicrystalline	Fixed tilt, 20 degree	Net-Metering

(*continued*)

TABLE 11.3. (*Continued*)

3) Valrico WRRF, Hillsborough County, Florida, USA	**WRRF Description**	The Valrico WRRF is permitted to treat an AADF of 45.4 ML/d (12.0 mgd) of wastewater. The facility is a Type I, oxidation ditch advanced wastewater treatment plant and includes a reclaimed water pumping station to serve the public access reuse system. Average power demand for the facility is 0.6 MWh/mo.				
	Drivers Photovoltaic for Generation	The county wanted to take advantage of their renewable sources of energy and the land available at the Valrico WRRF to diversify energy inputs, minimize energy cost, and become more sustainable				
	Photovoltaic Energy Generation Output	**Area of Panels**	**Types of Photovoltaic Panels**	**Types of Mounting**	**Types of Metering**	**Ground Coverage Ratio**
	2 MW alternating current (STC) proposed	12 Ac	Monocrystalline	Fixed tilt, 20 degree	Net-Metering	30%
4) Venice Beach WRRF, Florida, USA	**WRRF Description**	The WRRF consists of a 30.2 ML/d (8.0 mgd) 3-mo average daily flow, Type I advanced wastewater treatment (five-stage Bardenpho) domestic wastewater treatment facility and reclaimed water distribution pumping system. Average power demand for the facility is 0.6 MWh/mo.				
	Drivers Photovoltaic for Generation	The WRRF wanted to take advantage of their renewable sources of energy and the land available at the WRRF to diversify energy inputs, minimize energy cost, and become more sustainable.				
	Photovoltaic Energy Generation Output	**Area of Panels**	**Types of Photovoltaic Panels**	**Types of Mounting**	**Types of Metering**	**Average Daytime Energy Use**
	2.0 MW alternating current (STC) proposed	4 Ac	Monocrystalline	Fixed tilt, 20 degree, floating	Net-metering	92% solar, 8% utility power

purchase agreements) can be used to help capitalize upon available tax credits and other exemptions. This can provide value to private developers who can potentially pass along a portion of the savings to the utility (e.g., WRRF) in the form of lower power purchase rates. It is important to recognize that power purchase agreements often last 15 years or longer.

There are different critical factors affecting the capital and operational cost during photovoltaic energy applications (Sunrun, n.d.). In addition to factors described above, the factors that should be evaluated when considering this technology in a WRRF are:

- Solar Irradiance. Irradiance is defined as the radiant flux (power) received by a surface per unit area. Solar irradiance is the most critical factor for photovoltaic panels viability as it determines the energy output.
- Availability of Net Metering. Net energy metering (NEM) or net metering, is a billing agreement between power utility customer and the local power utility provider. It allows the customer to receive credits when the customers/utility produces excess electricity from "energy production facility" that is exported to the power grid. There are typically limits imposed regarding the maximum price that will be paid for the power exported to the grid.
- Land Availability. Available area for deployment of solar panels is a factor in the amount of solar photovoltaic panels that can be installed.

2.2 Thermal Energy

2.2.1 *Introduction*

Thermal energy is an important source of energy in wastewater, and recently, recovery of that energy has attracted more attention. This section describes the technology available for recovery of thermal energy, the application of thermal energy in WRRF, current practices, and several case studies.

2.2.2 *Sources of Thermal Energy*

There are three main sources of heat that may be available at a WRRF: (1) heat recovery through extraction of heat from wastewater, (2) recovery of waste heat, and (3) heat generation through solar devices.

2.2.2.1 *Heat Content of Wastewater*

Extracting heat from wastewater involves heat exchangers, thermal fluid, and heat pumps. Figure 11.3 presents a schematic of one approach to transfer heat from the wastewater to a point of use. Wastewater flows through a heat exchanger in which heat is transferred to a thermal fluid that is at

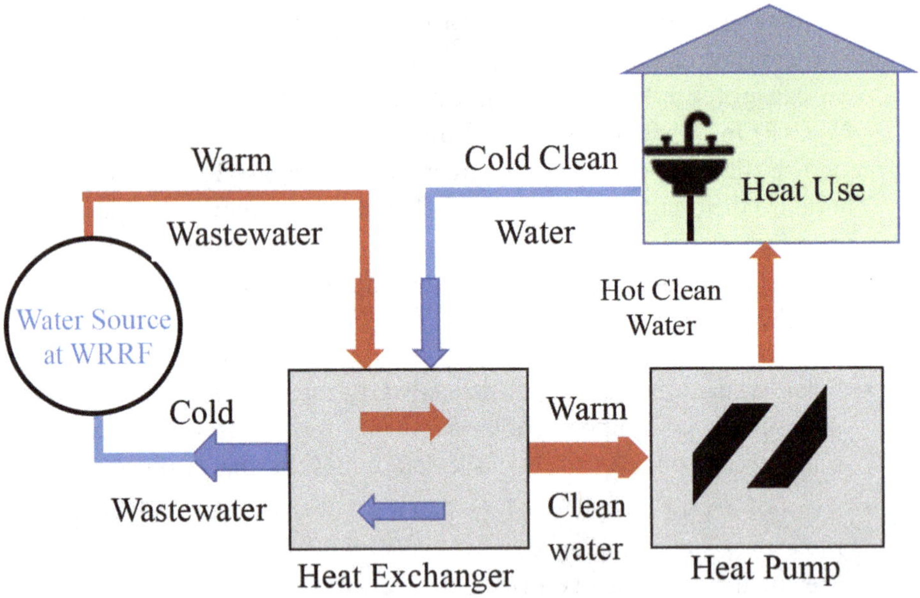

FIGURE 11.3. Heat recovery process

a lower temperature. Heat is recovered by an evaporator, which is similar to the heat exchanger. In this system, the compressor moves the fluid and raises the pressure and temperature of the working fluid. Then the compressed fluid is led to a condenser where heat is rejected for usage. After the condenser, the cooled fluid flows to an expansion valve where the pressure and temperature of the working fluid are decreased to a level that heat can be transferred from wastewater to the fluid again. In the heat pump, the energy output is considerably higher compared to the electricity consumption of the compressor. It also has an advantage of being able to operate the other way around, that means producing cooling energy (Meggers & Leibundgut, 2011).

2.2.2.2 Waste Heat

Waste heat is produced in thermal processes. Sources of waste heat generally include hot combustion gases discharged to the atmosphere, anaerobic digestion, heated products exiting wastewater treatment processes, and heat transfer from hot equipment surfaces. The hot air or water flows through a heat exchanger and passes its heat to another fluid or operating area. Specifics of the air and water (e.g., amount, flowrate, size and capacity of heat exchanger, etc.) depend on the source of energy, how the recovered heat will be used, and the holding time.

2.2.2.3 Geothermal

Geothermal power is considered to be a sustainable and renewable source of energy for a WRRF. In geothermal process, the heat extraction is small compared with the Earth's heat content. As a source of renewable energy for wastewater treatment, geothermal has the potential to meet a good percentage of the treatment facility's own energy demand. The amount of energy available from the geothermal system depends on the flowrate of the fluid and the difference between geothermal supply temperature and the geothermal reinjection temperature.

In this technology, the wastewater absorbs warmth from underground during the winter, and the pump brings this heat to the building above. In the summer, heat pumps can run in reverse and help cool buildings. The geothermal system is generally designed oversized because the values for the capacity and the energy supplied by this technology is approximate and, therefore, it is sometimes difficult to come to a conclusion about the exact number of units installed (Lund, 2003). In the United States, geothermal heat pumps are sized for peak cooling load. In contrast, most units in Europe are sized for the heating load and are typically designed to supply the peak base load. Therefore, the units there can be in operation much higher full load heating per year (Lund & Boyd, 2016).

2.2.3 Uses for Recovered Heat in Water Resource Recovery Facilities

The recovered thermal energy could be a considerable source to fulfill energy demand at a WRRF. Recovered heat could be used for warming the buildings. Sludge treatment processes, such as anaerobic digestion and sludge drying, could also benefit from the recovered heat. Furthermore, excess heat can be transferred to the adjacent consumers situated outside the premises of the WRRF.

Studies found that the average annual wastewater temperature varies between 30 and 35 °C at the point of discharge (Brdjanovic et al., 1997). However, this is the temperature that wastewater characterizes at the time of their formation. But when the wastewater comes to the WRRF, its temperature decreases. This temperature drop depends on the size of the sewer system and the flowrate. The average temperature that wastewater as an inflow to the WRRF varies between 11 and 12 °C (Gilbert et al., 2014). This applies to all types of wastewater, such as gray water (generated during the use of showers, bathtubs, and washing machines) as well as municipal and industrial wastewater. After that, heat exchangers and heat pumps collect the wastewater and recover the thermal energy from the system. The mechanism for converting this energy takes place in a multiphase system in wastewater and in systems inside the heat pump system.

2.2.4 Current Practice of Thermal Energy Implementation

In the United States, recovery of thermal energy typically occurs in a process used for biogas utilization or recovery of waste heat from biosolids incineration. The recovery of the thermal heat from wastewater by heat exchangers and heat pumps is not common. The application of this technology is still limited to few countries with approximately 500 applications worldwide (Mo & Zhang, 2013), primarily practiced in Switzerland, Germany, and Scandinavia (Hepbasli et al., 2014). In Switzerland, cantons and municipalities can draw on cantonal and municipal energy structure plans that analyze the energy supply and identify strategies to optimize the use of localized waste heat and renewable energy sources (Cajot et al., 2017). Guidance on such energy structure plans prioritizing the use of energy in a ranking list in which the top priority is given to localized high-grade waste heat followed by localized low-grade waste and ambient heat (e.g., from WRRFs). In Austria, wastewater thermal recovery has been implemented recently (Kretschmer et al., 2016).

2.2.5 Barriers and Incentives for Implementing Thermal Energy at Water Resource Recovery Facilities

The thermal energy available in wastewater is considered as low-grade heat that can be used for both cooling and heating purposes. There are few barriers that cause this energy to be under utilized and ultimately decreases the efficiency. These barriers include the uncertainty of lowering wastewater temperatures on the sewer network, risks of increased blockages, reduced sewer capacity, risk of damage to sewer pipes, and effects to WRRF efficiency. Moreover, if the hot water consumption goes down, then the benefit from the WRRF is decreased. Achieving environmental certifications to install this thermal recovery system is also challenging in a few countries. Along with that, the financial aspect of the system is important, including the comparison of space, the energy savings, investment costs, and so forth. Wastewater thermal energy recovery also faces practical challenges such as: the need for heat pumps and heat exchangers; the comparatively low cost of natural gas; longer payback period of heat recovery systems; and, in some cases, finding a suitable application for the low-grade heat. Moreover, some heat exchangers are only suitable for grey water and not for wastewater, which require a source separated wastewater system.

2.2.6 Case Studies

This section covers few case studies of full-scale municipal WRRFs that are currently practicing thermal energy generation at their facility. The following

WRRFs will be covered where three of them are in the United States and the other two are in Europe.

- The Southeast Water Pollution Control Plant, Philadelphia
- James C. Kirie Water Reclamation Plant, Chicago
- Heat recovery plant in Hammarbyverket, Stockholm, Sweden
- Heat recovery plant in Sandvika, Oslo, Norway
- Albert Lea WRRF, Minnesota

2.2.6.1 The Southeast Water Pollution Control Plant, Philadelphia

The Philadelphia Water Department installed the geothermal system at the Southeast Water Pollution Control Plant, which is the first geothermal facility in the United States that provides building heat using domestic wastewater. This facility consists of a 1 million Btu/hr unit located in a basement and heat is directly accessed from the adjacent wastewater channel. To make the best use of the existing piping system, the wastewater unit is tied into the facility's heat loop ahead of the boilers, so the boiler controls are still in operation. When the heat pump produces enough water at 54.4 °C (130 °F), the boilers remain off. If the load exceeds the capacity of the heat pump, then the boilers respond. Figure 11.4 illustrates the schematic of thermal heating and cooling system of the facility.

This geothermal installation consists of a filter, heat pump, and the appurtenant piping needed to extract the thermal energy from the wastewater arriving at the facility and to use this energy to heat the facility's compressor building and gallery space. A portion of the wastewater entering the facility passes through an antiblock filter. This filter handles up to 1003 L (265 gal) of wastewater per minute and removes wastewater solids that would otherwise interfere with the heat pump downstream. The highly efficient heat pump extracts up to 1 031 790 kJ (978 000 Btu) of energy per hour from the wastewater. The heat pump generates four units of heat for every unit of electric power supplied, which significantly reduces the natural gas required for heating the building. The wastewater geothermal system fulfills approximately 20% of the facility's heat load, which translates to a projected $15,000 in annual energy savings.

2.2.6.2 James C. Kirie Water Reclamation Plant, Chicago

A "sewer thermal" heating and cooling system was installed at the James C. Kirie Water Reclamation Plant in Des Plaines, Illinois. The facility has a daily average flow of 196 ML/d (52 mgd) with a capacity of 378 ML/d (100 mgd) and serve approximately 217,000 people living in a 168.9 km^2

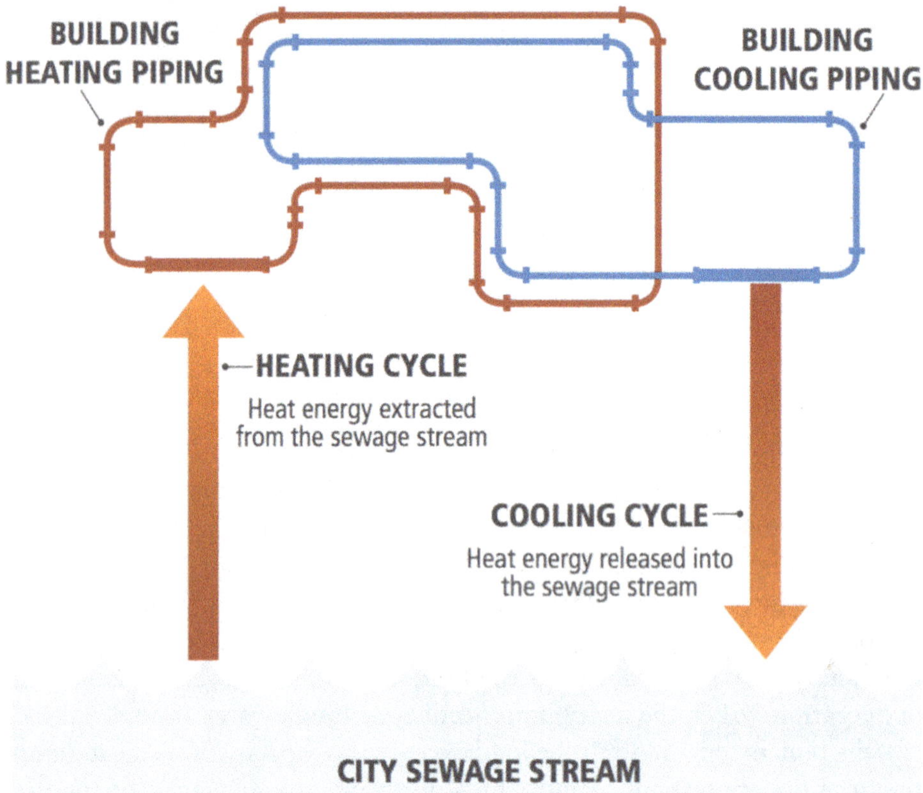

FIGURE 11.4. A schematic of wastewater heating and cooling of geothermal system at the Southeast Water Pollution Control Plant, Philadelphia (adapted from the Philadelphia Water Department).

(65.2 sq mi) area. In the installed thermal system, heat pumps collect heat during winter through fluid circulating in pipes called loops, which are placed below ground or in a body of water. This circulating fluid carries heat indoors. However, in summer, the loop draws away indoor heat and carries it underground or in the water, where it is absorbed. The system has both closed and open loops. In a closed loop system, the circulating water stays within the pipes. In an open loop, effluent water is used directly instead of the circulating fluid and is pumped in and out of the system. The heat pump system in this facility reduces energy consumption and corresponding emissions up to 44% compared to air-source heat pumps and up to 72% compared to electric resistance heating with standard air-conditioning equipment. It serves the 929 m^2 (10,000 sq ft) administration building and reduces the building's heating and air conditioning costs by 50%.

2.2.6.3 Heat Recovery Plant in Hammarbyverket, Stockholm, Sweden

Hammarbyverket is the world's largest heat pump facility that receives a wastewater discharge of 94 to 435 ML/d (25 to 115 mgd) and extracts heat from purified wastewater. It is located in the southern part of Stockholm and has been providing households with heat since 1986. The facility has installed seven heat pumps with the total capacity of 225 MW. The installed heat pumps produce approximately 1235 GWh which is enough to supply 95,000 two-bedroom apartments with heat. In addition, a 315-kW turbine had been installed to produce electricity from the hot side stream. The temperature of the incoming wastewater from the facility in Hammarbyverket varies from 7 to 22 °C. After the heat pump system, the temperature of the effluent is 1 to 5 °C. This heat pump system can raise the temperature up to 80 °C. However, two electric and two bio-oil boilers had been installed to increase the temperature up to 120 °C during the time of peak consumption and colder periods (Fortum, 2013).

2.2.6.4 Heat Recovery Plant in Sandvika, Oslo, Norway

The heat recovery facility in Sandvika, Norway, uses a heat exchanger in the main sewer. Along with that, a heat pump system has also been installed to get the maximum benefit from the system. In this system, three heat pumps with a capacity of 20 MW for heating and 18 MW for cooling energy is used to treat the wastewater before the heat is recovered. The average wastewater discharge is 215 ML/d (57 mgd) and the temperature of wastewater is allowed to drop to 4 °C. The recovered heat from this system supplies energy to the offices and residential buildings, which actually meet the demand of more than 50% of their energy consumption (Vestberg, 2017).

2.2.6.5 Albert Lea WRRF, Minnesota

With more than two-thirds of the up-front installation and maintenance costs covered by the State of Minnesota and a local utility, the Albert Lea WRRF in Albert Lea, Minnesota, developed a 120-kW heating system, which saves the facility approximately $100,000 in annual energy costs. Approximately 70% of the savings resulted from reduced electricity and fuel purchases, and the remainder from reduced maintenance costs. The heating system is fueled by biogas produced in the anaerobic digester. The biogas is first conditioned to remove impurities, leaving methane to be burned in the microturbines. This is accomplished by sending the gas through a suction scrubber, a scroll gas compressor, carbon filter vessels for siloxane removal, and then cooling the gas to remove moisture.

2.2.7 Cost–Benefit Analysis

The capital cost for installing thermal recovery facilities in WRRF includes heat pump cost, heat exchanger cost, structural cost, and piping and fittings cost. The operational and maintenance cost also cause a significant addition to the overall cost of the facilities. However, considering the energy savings and energy recovery from it, the technology itself has a significant benefit. A U.S. Environmental Protection Agency (U.S. EPA) study showed that a 10% reduction in the energy use of a WRRF would collectively save approximately $400 million and 5 billion kWh annually (U.S. EPA, 2013a). The key to the success of a thermal recovery project that requires electricity to power a heat pump is the spread in cost per British thermal unit (Btu) between the price of electricity and the unit price of natural gas, or the fuel used to generate heat. The cost per kilowatt-hour for electricity is converted to the equivalent cost per Btu. This value is compared to the cost per Btu of the fuel used to generate heat. The cost per Btu of electricity must be less than the cost per Btu of the heating fuel. But the margin between cost per Btu of electricity and cost per Btu of the heating fuel has to be large enough to justify the investment and to fund the capital investment in equipment and associated operating and maintenance costs.

2.3 Hydraulic Energy

2.3.1 Introduction

Hydraulic energy may be explored as an alternative source of renewable energy. The major benefit of using hydraulic energy in a WRRF is that it is a predictable and reliable source of energy.

2.3.2 Technology and Energy Generation Mechanism

Hydraulic energy is generated with the kinetic energy of moving water. A hydroelectric turbine transfers kinetic energy from moving water to mechanical energy that powers an electrical generator that creates the electricity.

Hydropower generation typically depends on two critical parameters: head (pressure) and water flow. Thus, any water infrastructure system that includes discharge and available head are potential sources of hydropower generation. In municipal wastewater systems, there are two potential locations where hydraulic energy can be generated, one is at influent location and other is at effluent location of WRRF.

There are two types of turbine technology: impulse and reaction type turbines (Choulot et al., 2012). Impulse turbines are applicable for high head and low flow conditions. This turbine uses the velocity of moving water that rotates the turbine blade and generates electricity. The reaction turbine is mostly suitable for low head and higher flow conditions and this turbine uses the pressure of moving water for power generation.

Table 11.3 presents the different types of turbine technology applicable for hydropower generation.

2.3.3 Hydraulic Energy as Form of Renewable Energy and Its Use in Water Resource Recovery Facilities

Installing a turbine upstream of the WRRF is presented in Figure 11.5a. In this approach, the raw wastewater from the collection system flows to a preliminary treatment area, which is equipped with a thin trash rack and rack cleaner/desilting chamber (Bousquet et al., 2017). The wastewater from this chamber discharges through penstock (a connecting pipe) to the hydropower plant (turbine) located immediately upstream of the WRRF. The electricity generation from raw wastewater flow requires frequent maintenance works

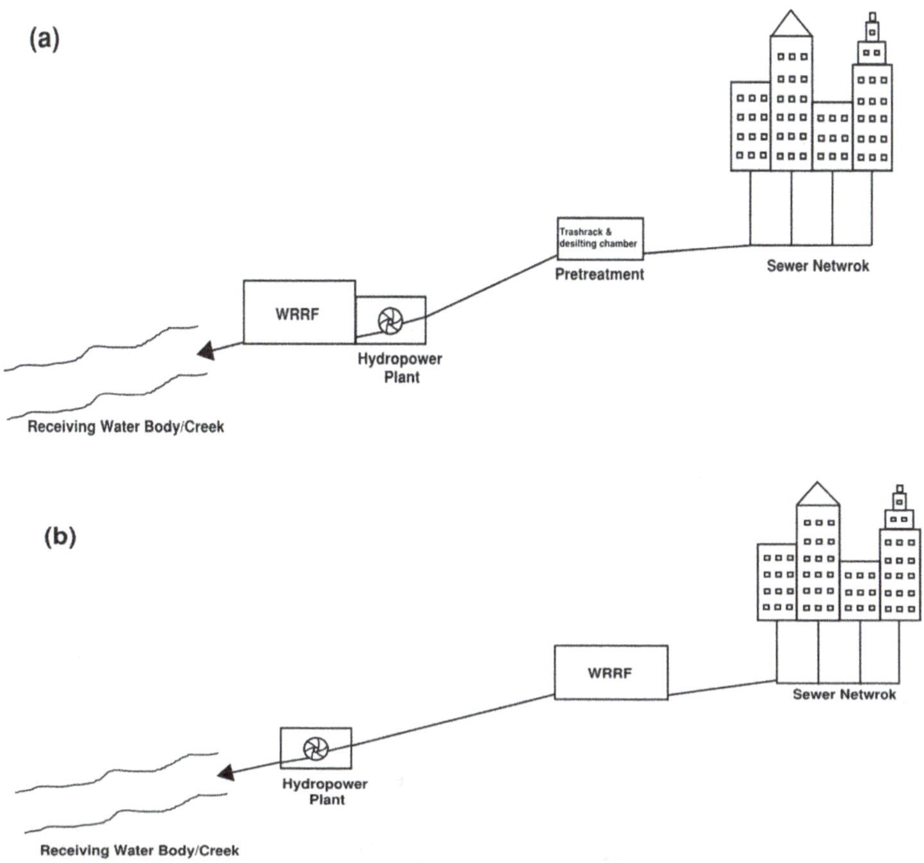

FIGURE 11.5. Hydropower plant operation in a WRRF at (a) upstream of raw wastewater location and (b) downstream of treated final effluent location

to prevent the mechanical failure of hydropower plant equipment. In addition, hydrogen sulfide corrosion can be an issue for equipment damage. In this approach, the head is available between the point where all sewer lines from a designated urban network is collected and the WRRFs influent location. For efficient electricity generation, the turbine must be installed as close to the inlet location of the WRRF to maximize the head.

The second approach is to generate electricity from hydropower by installing a turbine before effluent outfall at the WRRF (Figure 11.5b). The treated effluent from the WRRF flows through a penstock to a turbine (or multiple turbines) before being discharged to outfall or waterbodies or creek. The hydropower generation from this approach is primarily dependent on topographic elevation difference (available head) between the WRRF's final effluent point and outfall elevation (Bousquet et al., 2017). Although hydropower can be generated through the turbine, a final effluent bypass line should be installed for the case of the turbine being out of service or maintenance. This way, final effluent can be discharged to the outfall without any operational interference of the WRRF.

Hydropower generation can be adjusted based on the effluent flow loading conditions and electricity consumption of the WRRF. The generated electricity from the hydropower plant is delivered to the WRRF through an individual transmission line that interconnects with the WRRF's electric distribution system. In addition, the hydropower plant can also be connected to the electric utility grid at the nearest access point.

2.3.4 State of the Practice and Development of Hydraulic Energy Implementation in Water Resource Recovery Facilities

Available head and discharge flow are two critical factors for implementing hydropower generation. Whereas available head is dependent on geographic and topographic locations, discharge flow is mostly dependent on population density and combined sewers systems (CSS) (Bousquet et al., 2017). Several WRRFs have practiced hydropower generation. The head ranges between 4 to 450 m (13 to 1476 ft) and discharge flow ranges between 0.1 to 26 m^3/s (3.53 to 918 cfs) (Bousquet et al., 2017). There are only four WRRFs located in United States that are generating electricity using hydraulic energy. In Europe, Switzerland is one of countries that has developed the most hydroelectric projects related to WRRFs. **Table 11.4 shows different types of turbine technologies.** Table 11.5 summarizes hydropower plant installation in WRRFs located around the world.

TABLE 11.4. Different types of turbine technologies (adapted from Capua et al., 2014).

Class	Type of Turbine	Flow Type	Head Type
Impulse	Pelton wheel	Low	High
	Cross flow	Low	High
Reaction	Francis	High	Low
	Kaplan	High	Low
	Bulb/Microturbine	High	Low
	Pump as turbine	High	Low
Miscellaneous	Gravitational water vortex flow power plant	Low	Low

TABLE 11.5. Summary of hydropower plant operation in WRRFs located around the world (adapted from Bousquet et al., 2017; Maktabifard et al., 2018).

Country	WRRF	Flow (m^3/s)	Head (m)	Technology	Installed Total Power (kW)
		Hydropower plant at downstream final effluent of WRRF			
USA	Deer Island, Boston, MA	2 × 13.1	8.8	2 Kaplan turbines	2000
	Point Loma, San Diego, CA	7.6	27	Francis turbine	1350
	Clark County, Las Vegas, NV	7.8	120	No information	9722
Switzerland	Aire, Geneva	3.2	5	Kaplan turbine	200
	Engelberg	0.2	54.4	Pelton turbine	50
	Grachen	0.1	365	Pelton turbine	262
	La Asse, Nyon	0.3	94.25	Pump as turbine	220
	La Douve I, Leysin	0.1	545	Pelton turbine	430
	La Douve II, Leysin	0.1	83	Pelton turbine	75
	Morgental, St. Gallen	0.8	190	Pelton turbine	1350

(*continued*)

TABLE 11.5. (*Continued*)

Country	WRRF	Flow (m^3/s)	Head (m)	Technology	Installed Total Power (kW)
Hydropower plant at downstream final effluent of WRRF					
Austria	Seefeld, Austria		625	Turbine	231
	Vienna WRRF, Austria	6.5	5	Vertical axis turbine	174
Germany	Emmerich, Germany	0.4	3.8	Archimedes screw	13
Australia	North Head, Sydney, Australia	3.5	60	Kaplan turbine	4500
Jordan	As Samra, Jordan	3.2	41	2 Francis turbines	1680
Hydropower plant at upstream sewer collection system before entering to WRRF					
Jordan	As Samra	3.2	104	2 Pelton turbines	1600
England	Elsholt	2.6	—	2 Archimedes screws	180
Switzerland	Profay, Le Chable	0.1	449	Pelton turbine	350

2.3.5 Barriers and Incentives for Implementing Hydraulic Energy in Water Resource Recovery Facilities

To implement a hydraulic energy generation project in a WRRF, there are two major barriers that need to be considered in the project:

1. Federal Energy Regulatory Commission (FERC) regulations require that all hydroelectric projects need to obtain the FERC permitting process.
2. The amount of power generation may not be significant compared to energy required for low flow capacity WRRF. For example, a 38 ML/d (10 mgd) WRRF with a 98-m (30-ft) drop at the effluent outfall can generate less than 2.5% of the total amount of energy required by the facility to operate.

The major limitations of implementing a hydropower plant in a WRRF for both raw wastewater (influent) and final effluent location include lack of available head, variation of flowrate, and failure of hydroturbine, especially at the raw wastewater location (McNabola et al., 2014). The chemical composition of raw wastewater has a negative effect on hydropower systems because of particular pH, oxygen content, and temperature, which can cause explosive and corrosive atmosphere (Berger et al., 2013). In addition, biological fouling or solids abrasion factors may have negative influence on mechanical equipment failure. Thus, all mechanical equipment, devices, and accessories associated with a hydropower plant require proper design aspects. However, the best practice to generate hydropower from a WRRF would be from effluent outfall location to avoid all complexity compared to hydropower generation at influent location of the WRRF.

2.3.6 Case Studies

This section focuses on full-scale municipal WRRFs located in the United States that are currently practicing hydropower generation at their facility. To the best of authors' knowledge, four facilities have a hydropower plant located at final effluent point to generate electricity:

- Deer Island Wastewater Treatment Facility, Boston, Massachusetts
- Point Loma Wastewater Treatment Plant, San Diego, California
- Clark County Water Reclamation District, Las Vegas, Nevada
- City of Warren Water Pollution Control Facility, Warren, Ohio

2.3.6.1 Deer Island Wastewater Treatment Facility, Boston, Massachusetts, USA

The Deer Island Wastewater Treatment Facility (1381 ML/d [~365 mgd]) located in Boston, Massachusetts, is one of the first WRRFs in the United States that generates electricity using hydraulic energy. It is producing approximately 26% of the facility's electricity through renewable sources such as hydropower, methane gas recovery, solar power, wind power, and steam turbine generation (Massachusetts Water Resources Authority [MWRA], 2013). The Deer Island facility has been generating electricity using hydraulic energy at the final effluent point since 2002. The facility is using two Kaplan turbines, each with 1 MW capacity. The hydropower plant in this facility produces more than 6 million kWh annually, saving the electricity cost approximately by $600,000 per year (Capua et al., 2014; MWRA, 2013). The Deer Island facility uses 70% of total electricity consumption purchased by the MWRA annually. The facility produces hydropower electricity by 5.1% of total electricity consumption in each year (Capua et al., 2014; MWRA, 2013).

2.3.6.2 Point Loma Wastewater Treatment Plant, San Diego, California

The Point Loma Wastewater Treatment Plant located in San Diego, California, has a permitted capacity of 908 ML/d (240 mgd), and the facility is situated on a bluff that is 295 m (90 ft) above the Pacific Ocean where the facility discharges primary treated effluent (City of San Diego Public Utilities Department, 2014). This facility has a unique advantage on final effluent discharge point to outfall compared to other treatment facilities. This is because it discharges its final effluent through a 7.2-km-long (4.5-mi-long) ocean outfall (City of San Diego Public Utilities Department, 2014; Capua et al., 2014). The outfall is one of the longest and deepest (94.5 m [~310 ft]) in the United States, providing the great opportunity for the hydropower electricity generation (City of San Diego Public Utilities Department, 2014; Capua et al., 2014). The facility uses Francis turbine at the final effluent point to generate up to 1.35 MW of electricity, which is sufficient to supply electricity to more than 10,000 homes. The generated electricity is then sold to San Diego Gas and Electric Company (U.S. EPA, 2013b).

2.3.6.3 Clark County Water Reclamation District, Las Vegas, Nevada

The Clark County Water Reclamation District with a capacity of 677 ML/d (179 mgd) captures hydraulic energy from the final effluent location. The available head between the final effluent point and the outfall is approximately 120 m (384 ft), and the outfall location is 22 km (~13.7 mi) away from the facility. A hydroelectric generator with a capacity of 16 MW is installed to produce hydropower, and the hydropower plant generates electricity over 85 million KWh/yr. The facility's required electricity consumption for pumping is approximately 4 million KWh/yr. The remaining 95% of electricity from the hydropower plant supplied to an electrical grid in the vicinity of the effluent outfall resulted in revenue for the facility (Karafa et al., 2007).

2.3.7 Cost–Benefit Analysis of Renewable Energy Use in Water Resource Recovery Facilities

There are some possible incentives that can be considered to promote renewable energy applications, such as tax incentives or subsidies, renewable energy funds, voluntary green schemes, green certificate schemes, and price-based instruments (Bousquet et al., 2017). Literature suggests that electricity generation using hydraulic energy in WRRFs would not be economically feasible if there are no government incentives (Gaius-obaseki, 2010). The potential savings on hydropower generation may vary facility to facility depending on geography and other variables, and the payback period may not be immediate. The capital cost for hydropower plant installation in a WRRF includes the cost of the hydroturbine, structural cost (concrete such as turbine foundation), piping and fittings cost, and contingency for construction (Capua et al., 2014).

Capua et al. (2014) reported that a reasonable payback period for hydropower plant should be between zero and 10 years. This can be considered as a quarter of the length of the expected life of the project equipment, and this will provide the benefit for the facility to avoid risk of premature failure of the equipment. In addition, MWRA (2014) and Capura et al. (2014) also stated that the service life for a hydropower plant should be 20 years at minimum to achieve economic feasibility of hydraulic energy utilization.

2.4 Wind Energy

2.4.1 Introduction

Wind turbines can be used as auxiliary and supplemental power sources for WRRFs (Spellman, 2013b). The use of wind power to generate electricity at WRRFs offers several advantages: (a) energy production can be inexpensive; (b) it is infinitely sustainable form of energy with no fuel requirement for operation; (c) it produces no harmful air, greenhouse gases, toxic/radioactive waste, or water pollution generation; and (d) leaves usable land below each turbine for other purposes including treatment units and more. Wind turbine capacity ranges from few kilowatts to more than 5 MW (Merchant, n.d.; U.S. EPA, 2013c). Wind energy can be a major source of renewable energy for WRRFs specially located in coastal areas, mountainous areas, and/or islands (Ali et al., 2018; Office of Energy Efficiency & Renewable Energy [EERE], n.d.-a; U.S. EPA, 2013c). With a steady wind flow, this could be sufficient to meet the electric energy needs of a 37.8 to 57.8 ML/d (10 to 15 mgd) WRRF. This section discusses the technology and mechanism to convert wind into energy, its uses as renewable energy in a WRRF, the state of its implementation, the barriers and incentives, and relevant case studies.

2.4.2 Technology and Energy Generation Mechanism

A wind turbine or windmill with a wind generator converts the energy in wind into mechanical energy first and then to electricity (U.S. EPA, 2013c). There are two types of wind turbines, vertical axis and horizontal axis. Horizontal axis wind turbines dominate most of the wind industry. *Horizontal axis* means the rotating axis of the wind turbine is horizontal, or parallel with the ground. They are primarily used in large wind application. However, vertical axis turbines have their place in small wind applications (Spellman, 2013b; U.S. EPA, 2013c; Windpower Engineering & Development [WED], 2009). Horizontal wind turbines can produce more electricity from a given amount of wind. However, they are generally heavier and do not produce well in turbulent wind conditions. Horizontal turbines have optimal performance when pointed directly into the wind. Furthermore, the operation must be monitored and the position of the horizontal turbine must be adjusted to achieve superior performance (Spellman, 2013b; WED, 2009). *Vertical axis turbines* are powered by wind coming from all 360-degrees directions, and even some turbines can be powered when the wind blows from top to bottom. Because of this versatility, vertical axis wind turbines are thought to be ideal for installations in which wind conditions are not consistent, or where as a result of public ordinances the turbine cannot be placed high enough to benefit from a steady wind. Wind power production occurs when wind speed > 11.3 kph (7 mph) and shuts down at speeds > 72.4 kph (45 mph) to protect the machinery of the windmill (U.S. EPA, 2013c).

2.4.3 Wind Energy Generation as Form of Renewable Energy and Its Uses in Water Resource Recovery Facilities

Several WRRFs throughout the United States have installed or are considering the installation of wind turbines (Spellman, 2013b). Wind energy can be effectively installed in WRRFs in areas with relatively high wind speed (11.3 to 72.4 kph [~7 to 45 mph]) to make the investment cost-effective (U.S. EPA, 2013c). Wind turbines exhibit many differences in their size. The size of the turbine typically depends on what its primary function is and where it is located. Small turbines are also used in conjunction with other systems such as photovoltaic systems or diesel generators (U.S. EPA, 2013c). Larger turbines above 100 kW and sometimes even megawatts are used in a "wind farm" type array, which groups together several of these turbines in a large open area that are all connected to an electrical grid. Wind power has become affordable enough to compete with fossil fuels (National Academics of Sciences Engineering Medicine [NAS Eng. Med.], 2007). In fact, many utility services around the world offer wind-generated electricity at a premium of 2 to 3 cents/kWh. On the other hand, coal-based energy is 4.8 to 5.5 cents/kWh or 11.1 to 14.5 cents/kWh for nuclear power (NAS Eng. Med., 2007; EERE, n.d.-b). Wind energy is less expensive than any other new source of electric generation except for natural gas. Many WRRFs can consider wind energy to complement their energy need from the main grid (Spellman, 2013b; U.S. EPA, 2013c).

2.4.4 Barriers and Incentives for Implementing Wind Energy at Water Resource Recovery Facilities

Different types of wind turbines, a mature technology, have been in operation in many countries across the globe for a long time, but growth of the installed wind power is hampered by a number of barriers. Some of those are applicable to WRRFs but the added benefits and incentives of wind power at WRRFs can make it a sustainable investment for long term (Ali et al., 2018; Spellman, 2013b). The detailed list of barriers and incentives in WRRF has been provided in Table 11.6.

Similar to solar photovoltaic, the use of a power purchase agreement with a third-party private developer may allow a utility to more effectively realize tax benefits offered by distributed renewable generation such as wind energy (EERE, 2020b). For successful implementation of wind energy at WRRF, the siting requires thoughtful planning, and the long-term needs should be considered.

2.4.5 Case Studies

There are several WRRFs in the US that have successfully installed wind turbines, as shown in Table 11.7.

TABLE 11.6. Barriers and incentives of wind energy in WRRFs (Ali et al., 2018; EERE, n.d.-b; Spellman, 2013b; The National Academy of Sciences Engineering Medicine, 2007).

Barriers	Incentives
Land requirements and relatively high cost of energy	Federal programs that provide direct cash to energy producers and consumers
Environmental effects and effects on natural habitat and wildlife	Tax expenditures that reduce tax liabilities of firms/utilities or individuals that produce, distribute, transmit energy
Public acceptance and technical challenges of project installation	Support for rural utilities
Grid interconnection	Loans and loan guarantees that provide support for energy technologies by guaranteeing the repayment of loans
Visual effect, audible noise, telecommunication interference	Usable land with no harmful effects on air, water, and waste

TABLE 11.7. WRRFs in the United States with successful wind turbine installations

1) Massachusetts Water Resources Authority (MWRA), Deer Island, Boston, Massachusetts	**WRRF Description**	The MWRA's Deer Island WRRF treats an average of 1381 ML/d (365 mgd) wastewater from 43 communities in greater Boston and is one of the largest electricity users in the Northeast. Deer Island currently self-generates 26% of its electricity needs and more than half of the island's energy demand is provided by on-site, renewable generation. Two 600-kW wind turbines (57.9 m [≈190 ft]) were installed and commissioned on Deer Island in November 2009.
	Wind Turbines Energy Generation	The wind turbines generate more than 2M kWh per year for an annual estimated savings of $250,000.
2) Atlantic County Utilities Authority (ACUA) Wastewater Treatment Facility, New Jersey	**WRRF Description**	ACUA's Wastewater Treatment Facility with a capacity of 151 ML/d (40 mgd) treats the wastewater of 14 county municipalities as well as sludge and septage generated throughout the county. The Jersey-Atlantic Wind Farm consists of five, 380-foot turbines. The power produced, in conjunction with solar energy project, is used to operate the WRRF, with any excess energy provided to the main power grid. The estimated cost of the 7.5-MW wind farm was $12.5 million with an expected cost saving of $350,000 (Spellman, 2013b).
	Wind Turbines Energy Generation	The 115.8-m (380-ft) turbines produce approximately 7.5 MW of power, which can power approximatively 2500 homes. The ACUA wind farm can supply more than 60% of the electricity required by the facility annually. The remaining electricity can be bought from the local power grid when windmills are not at peak capacity (during calm or gusty weather). The cost of wind-generated electricity is 7.9 cents/kWh delivered which cheaper than main grid electricity.

3) Browning, Montana	**WRRF Description**	In late 1999, the Town of Browning, Montana, and the Blackfeet Nation installed four Bergey Excel 10-kW wind turbines adjacent to the Town's WRRF. The turbines provided approximately one-quarter of the facility's electricity and displaced energy bought from the main grid.
	Wind Turbines Energy Generation	This project is subsequent to the Blackfeet 100-kW Wind Turbine Demonstration Project. Currently, the Blackfeet have an agreement with Sea West Wind Power for developing an additional 22.8-MW wind farm on tribal land.
4) Lynn Wastewater Treatment Plant, Massachusetts	**WRRF Description**	The Lynn WWTP is a primary and secondary treatment plant. The facility can process 97.7 ML/d (25.8 mgd) flow during dry weather periods and > 419.4 ML/d (110.8 mgd) during wet weather conditions generated from Lynn, Swampscott, Saugus, and Nahant. Known as "the city of the wind," the city of Lynn constructed a 77.4-m (254-ft) high wind turbine that will save the city approximately $60,000 a year in electrical costs.
	Wind Turbines Energy Generation	The turbine can power approximately 10% of a 40-ac WRRF, cost $1.8 million to purchase. The Lynn turbine can provide approximately 1 million kW/yr of energy to the facility.

3.0 REFERENCES

Ali, A., Tufa, R. A., Macedonio, F., Curcio, E., & Enrico D. (2018). Membrane technology in renewable-energy-driven desalination. *Renewable and Sustainable Energy Reviews, 8(1),* 1–21.

Berger, V., Niemann, A., Frehmann, T., & Brockmann, H. (2013). Advanced energy recovery strategies for wastewater treatment plants and sewer systems using small hydropower. *Water Utility Journal*, *5*, 15–24.

Bousquet, C., Samora, I., Manso, P., Rossi, L., Heller, P., & Schleiss, A. J. (2017). Assessment of hydropower potential in wastewater systems and application to Switzerland. *Renewable Energy*, *113*, 64–73.

Brdjanovic, D., Van Loosdrecht, M. C. M., Hooijmans, C. M., Alaerts, G. J., & Heijnen, J. J. (1997). Temperature effects on physiology of biological phosphorus removal. *Journal of Environmental Engineering*. https://doi.org/10.1061/(asce)0733-9372(1997)123:2 (144)

Cajot, S., Peter, M., Bahu, J. M., Guignet, F., Koch, A., & Maréchal, F. (2017). Obstacles in energy planning at the urban scale. *Sustainable Cities and Society*. https://doi.org/10.1016/j.scs.2017.02.003

Capua, M. R., Harris, C. D., & Dzwonkoski, J. (2014). *Reclamation of Power in Wastewater Treatment Facilities*. Digital WPI.

Choulot, A., Denis, V., & Punys, P. (2012). Integration of small hydro turbines into existing water infrastructures. *InTech*, 239–276.

City of San Diego Public Utilities Department. (2014). *Point Loma wastewater treatment plant secondary equivalency* [Fact sheet]. Pure Water San Diego Program.

Fortum. (2013). Hammarbyverket. Retrieved March 5, 2020, from https://www.stockholmexergi.se/om-stockholm-exergi/vara-anlaggningar/hammarbyverket/

Gaius-obaseki, T. (2010). Hydropower opportunities in the water industry. *International Journal of Environmental Sciences*, *1*(3), 392–402.

Gikas, P. (2017) Towards energy positive wastewater treatment plants. *Journal of Environmental Management, 203,* 621–629.

Gilbert, E. M., Agrawal, S., Karst, S. M., Horn, H., Nielsen, P. H., & Lackner, S. (2014). Low temperature partial nitritation/anammox in a moving bed biofilm reactor treating low strength wastewater. *Environmental Science and Technology*. https://doi.org/10.1021/es501649m

Hepbasli, A., Biyik, E., Ekren, O., Gunerhan, H., & Araz, M. (2014). A key review of wastewater source heat pump (WWSHP) systems. *Energy Conversion and Management*. https://doi.org/10.1016/j.enconman.2014.08.065

Karafa, D., Devlin, J., Yankovich, D., & Froehlich, D. (2007). Effluent to energy and it is renewable. In: *Proceedings of the Water Environment Federation*, WEFTEC, 4563–4577.

Kretschmer, F., Neugebauer, G., Kollmann, R., Eder, M., Zach, F., Zottl, A., Narodoslawsky, M., Stoeglehner, G., & Ertl, T. (2016). Resource recovery from wastewater in Austria: Wastewater treatment plants as regional energy cells. *Journal of Water Reuse and Desalination*. https://doi.org/10.2166/wrd.2015.119

Local Government Commission. (n.d.) Retrieved October 2020 from https://lgc.org/wordpress/docs/seec/lgpm/cca_lgpm-Howard-Brewen-101014.pdf

Lund, J. W. (2003). Direct-use of geothermal energy in the USA. *Applied Energy*. https://doi.org/10.1016/S0306-2619(02)00129-0

Lund, J. W., & Boyd, T. L. (2016). Direct utilization of geothermal energy 2015 worldwide review. *Geothermics*. https://doi.org/10.1016/j.geothermics.2015.11.004

Maktabifard, M., Zaborowska, E., & Makinia, J. (2018). Achieving energy neutrality in wastewater treatment plants through energy savings and enhancing renewable energy production. *Reviews in Environmental Science and Bio/Technology*, *17*(4), 655–689.

McNabola, A., Coughlan, P., & Williams, A. P. (2014). Energy recovery in the water industry: An assessment of the potential of micro-hydropower. *Water and Environment Journal*, *28*(2), 294–304.

Meggers, F., & Leibundgut, H. (2011). The potential of wastewater heat and exergy: Decentralized high-temperature recovery with a heat pump. *Energy and Buildings*. https://doi.org/10.1016/j.enbuild.2010.12.008

Merchant, B. (n.d.). *In the future, will wind turbines be everywhere?* Retrieved October 2020 from https://science.howstuffworks.com/environmental/energy/future-wind-turbine.htm

Mo, W., & Zhang, Q. (2013). Energy-nutrients-water nexus: Integrated resource recovery in municipal wastewater treatment plants. *Journal of Environmental Management*. https://doi.org/10.1016/j.jenvman.2013.05.007

Massachusetts Water Resources Authority. (2013). *Renewable and sustainable energy initiatives at Deer Island*. Massachusetts Water Resources Authority, Boston.

Massachusetts Water Resources Authority. (2014). *Appendix 10: Expected useful life of capital projects*. Massachusetts Water Resources Authority, Boston.

Office of Energy Efficiency & Renewable Energy (n.d.-a). *Offshore wind research and development*. Retrieved October 2020 from https://www.energy.gov/eere/wind/offshore-wind-research-and-development

Office of Energy Efficiency & Renewable Energy (n.d.-b). *Wind energy financial incentives*. Retrieved October 2020 from https://windexchange.energy.gov/projects/incentives

Spellman, F. R. (2013a). *Solar power in water and wastewater infrastructure-energy efficiency and sustainability*. Taylor & Francis Group, CRC Press.

Spellman, F. R. (2013b). *Wind power in water and wastewater infrastructure-energy efficiency and sustainability*. Taylor & Francis Group, CRC Press.

Strazzabosco, A., Kenway, S. J., & Lant P. A. (2019). Solar PV adoption in wastewater treatment plants: A review of practice in California. *Journal of Environmental Management, 248*, 109337.

Solar Energy Industries Association (n.d.). *Photovoltaics*. Retrieved October 2020 from https://www.seia.org/initiatives/photovoltaics

Sunrun Inc. (n.d.). *Economics of solar energy*. Retrieved October 2020 from https://www.sunrun.com/solar-lease/cost-of-solar/solar-energy-economics

The National Academies of Sciences Engineering Medicine. (2007). *Environmental impacts of wind energy project*. Retrieved October 2020 from https://www.nap.edu/read/11935/chapter/8

U.S. Environmental Protection Agency. (2013a). *Energy efficiency in water and wastewater facilities*.

U.S. Environmental Protection Agency. (2013b). *Renewable energy fact sheet: Low-head hydropower from wastewater.*

U.S. Environmental Protection Agency. (2013c). *Wind turbines renewable energy fact sheet*. Retrieved October 2020 from https://www.epa.gov/sites/production/files/2019-08/documents/wind_turbines_fact_sheet_p100il8k.pdf

Vestberg, O. (2017). *Heat recovery from untreated wastewater. A case study of heat recovery from sewer line to district heating network*.

Windpower Engineering & Development. (2009). *Vertical axis wind turbines vs. horizontal axis wind turbines*. Retrieved October 2020 from https://www.windpowerengineering.com/vertical-axis-wind-turbines-vs-horizontal-axis-wind-turbines/

4.0 SUGGESTED READINGS

Marcelino, R. B. P., Queiroz, M. T. A., Amorim, C. C., Leão, M. M. D., & Brites-Nóbrega, F. F. (2015). Solar energy for wastewater treatment: Review of international technologies and their applicability in Brazil. *Environmental Science and Pollution Research, 22*, 762–773. DOI 10.1007/s11356-014-3033-2

National Renewable Energy Laboratory. (n.d.). Home Page. Retrieved October 2020 from https://www.nrel.gov/

New York State Energy Research and Development Authority. (2009). *Renewable energy recovery opportunities from domestic wastewater.*

Pakenas, L. J. (1995). Energy efficiency in municipal wastewater treatment plants—Technology assessment.

Power, C., McNabola, A., & Coughlan, P. (2014). Development of an evaluation method for hydropower energy recovery in wastewater treatment plants: Case studies in Ireland and the UK. *Sustainable Energy Technologies and Assessments*, 7, 166–177.

Torrey, D. A. (2011). *Hydropower from wastewater* (No. DOE/DE/EE0000403). Advanced Energy Conversion, LLC.

Yüksel, I. (2010). Hydropower for sustainable water and energy development. *Renewable and Sustainable Energy Reviews*, *14*(1), 462–469.

12

Looking Forward

John Norton, PhD, PE; Peter V. Cavagnaro, PE, BCEE, PMP; Tony Cecchini; Vikram M. Pattarkine, PhD, BCEEM; & Wendy Barrott

1.0 INTRODUCTION

The future of energy management at water resource recovery facilities (WRRFs) is a combination of

- energy management—use of control strategies to minimize the cost and use of energy;

- energy efficiency—improvement of existing unit processes and associated equipment;
- energy generation and recovery—the generation of energy from biosolids, wind, heat, and solar, and the recovery of energy from either heat or kinetic flow; and
- new treatment approaches—the avoidance of energy use through modification, or entire removal, of treatment components.

The purpose of this chapter is to briefly describe new and evolving technologies and best practices for energy management at WRRFs.

2.0 ENERGY MANAGEMENT

The cost of electric energy for most WRRFs is based on a combination of demand and consumption. Demand is the peak kilowatts used at any specific point in time, and consumption is measured as the kilowatts being used over a period of time, typically as kilowatt per hour (kWh). The total energy cost for a billing cycle is composed on the measure of those two components over a billing period, where the demand cost is the peak kilowatts used over the billing period times a set cost per kilowatt, and consumption is the total kilowatt hours used during the billing period, times the price per kilowatt hour. There can also be even more detailed aspects of both demand and consumption. For example, demand may be a function of both demand over the billing period and a fraction of peak demand over a longer period of time, for example, 12 months, 18 months, etc. The consumption cost may also vary based on time of day or time of week, and in other instances, might be based on a continuous real-time pricing structure that varies depending on the market.

2.1 Tariff Agreements

Often, utilities learn that utility tariff structures can inhibit efforts to conserve energy or produce energy on-site by either minimizing the benefit of reducing consumption or by placing undue financial and technical burden on self-generation.

Tariffs can minimize the benefit of energy conservation by shifting costs from a "consumption only" model toward a blended charge of both consumption and maximum demand over a billing cycle. In recent years, power companies have specifically addressed energy conservation measures by developing and advocating tariff structures with significant demand charges. In many cases, a water or wastewater utility demand charge can match, or even exceed, the utility's consumption charge.

Tariffs can also inhibit the integration of energy-generating processes by not recognizing the likely reduction in utility demand or by requiring extensive processing and control if instantaneous production exceeds the utility demand. For instance, energy recovery from high-pressure pumped water may be allowed, but energy generation greater than the peak energy usage at the specific facility recovering the energy may be required to be wasted.

2.2 Optimizing Operations Within Your Tariff

There is an almost universal issue between operational efforts and the local energy provider's tariff structure. Specifically, there is often divergence between those who pay the bills and those who operate and maintain the equipment, and often these two groups do not effectively communicate with each other. As a result, operations often do not realize the effects of their specific activities on overall facility energy costs, let alone receive any sort of guidance on how to modify their efforts so as to reduce costs.

Often, when energy managers analyze their energy bills, they find that long-going operational protocols and maintenance procedures have cost the utility thousands of dollars per year. The authors recommend turning the story around when presenting to operations and pointing out the opportunities for improvement.

As described in Chapter 2, knowing and understanding how the electric energy bill is calculated is fundamental to developing a successful strategy for managing energy costs. The first step is data mining of electric utility bills for the preceding several years. The extent of data mining will depend on the extent of site-specific changes to the utility's energy use but should go back at least 2 or 3 years so seasonal effects can be investigated. The second step is to contact the utility representative for the facility to discuss the rate structure, availability of alternative rates, and other information available to assist with managing electric use at the site. The next step is to analyze and investigate local energy use and corresponding costs, and, finally, to develop a management plan going forward.

It should be noted that many facility pumping stations, and occasionally even whole facilities, have neither process-level flow metering nor electrical metering, and thus it can be difficult to quantify the energy use reduction because of unit or process-level improvements.

2.3 Power Quality Management

In contrast to energy management, which focuses on long-term energy use and energy use patterns, power quality management emphasizes the role of extremely short-duration fluctuations in the power grid and the corresponding short and long-term effects on utility operations. These short-term power quality issues can cause facility issues, such as electrical panel trips and motor stoppages, and long-term damage accumulation resulting in increased electrical usage and reduced operational life. The main power quality issues are the following:

- voltage sag (or dip),
- short interruptions,
- long interruptions,
- voltage spike,
- voltage swell,
- harmonic distortion,
- voltage fluctuation,
- noise, and
- voltage unbalance.

Power quality metering is much shorter in duration than energy use metering. Power quality issues often last fractions of a second and must be metered with more precise equipment than a typical energy meter. As a result, power quality metering is often performed using temporary meter installations that capture a few days or weeks of data at a time.

Also of concern is the need for installation, operations, and maintenance of uninterruptible power supplies (UPS). Further, the UPS equipment provided by some of the major manufacturers can drive some of the power quality issues.

Because of the metering requirements, most utilities have barely even scratched the surface of power quality issues and power quality management.

2.4 Building Automation Systems

Building uses can account for 5% to 15% of the energy consumption at a typical WRRF. A building automation system (BAS) monitors heating, ventilation, and air conditioning; lighting; and other building functions and allows a central control to establish operational setpoints in much the same manner as with operational process controls. The operator can set temperature and lighting levels and schedules for individual spaces throughout the facility. It is a way to ensure the lights are on when people are working in a space and are turned off when not needed. Facility-wide security, such as video cameras and employee location services, can also be incorporated into a BAS. It should be noted that BAS should not be used for treatment process controls.

3.0 ENERGY EFFICIENCY

This section addresses the improvement of existing unit processes and associated equipment; for example, improved design of mixing or aeration systems.

Owners and operators face the ongoing question of whether to maintain and repair existing equipment or to replace and upgrade. In the short run, and using simple payback as the criteria, "maintain and repair" often wins. Thus, installed technology is not changed and can potentially become outdated.

The "replace and upgrade" option can come with an effect to marginal cost. For example, the decision might be to rewind a burned-out motor, replace with a standard efficiency motor, or replace with a high-efficiency motor. In the short term, rewinding might cost less and is common practice. However, this approach will not lead to an improvement in overall facility energy efficiency. However, by evaluating the longer time horizon of "duration that location will have a pump," the calculated investment in the higher cost but higher efficiency motor will often result in lower overall costs.

3.1 Evolving Technologies—Nanobubble Aeration

Nanobubble aeration seeks to improve oxygen transfer efficiency by reducing the size of the air bubble. Some of the claims for nanobubble aeration are neutral buoyancy resulting in a longer residence time within the fluid and high oxygen transfer efficiency (Xiao & Xu, 2020). Products are offered for flotation, odor control, supplemental aeration, and membrane bioreactor. The manufacturers offer various equipment configurations delivering 25 to 1000 gpm (Moleaer and $NanO_2$). Nanobubble aeration has been noted to inhibit the growth of aqueous plants, so care should be taken in its application (Wang et al., 2020).

3.2 Evolving Technology—Pumping

Pumping is one of the major consumers of electricity at WRRFs. Pumps are used to lift wastewater, feed chemicals, feed sludge to thickening and dewatering equipment, and transport reclaimed water.

Recent advances claimed by manufacturers include alternative impeller designs for both general and specific pumping applications. It may be difficult for a designer to specify alternative equipment as the result of established design criteria. Owners and operators considering these technologies for replacing or upgrading equipment should engage the respective vendors to pursue testing and validation of the technology claims.

3.3 Evolving Technology—Lighting

Recent years have seen huge increases in lighting efficiency because of a succession of new technologies, including light-emitting diodes (LEDs), integration with building automation systems, and improved lighting controls (Uddin et al., 2011). The improvements include increased design life, improved lighting aesthetics, and reduced energy costs. Although initially high, the capital cost of new lighting installations has also dropped, and the overall payback of lighting projects can range from 3 to 7 years.

Lighting projects can also improve workplace safety, especially in areas where critical work might be performed or expected.

4.0 ENERGY GENERATION AND RECOVERY

This section introduces the generation of energy from biosolids and the recovery of energy from either heat or kinetic flow. The emerging field of carbon redirection is included in Section 5, New Treatment Approaches.

There are well-established methods using anaerobic digestion with subsequent capture and use of the biogas and incineration with possible heat recovery. There are several emerging technologies that also use the chemical energy of biosolids for energy generation. These fall into two major categories, dry and wet, which refer to the starting water content of the organic biomass (Basu, 2013). The dry processes include gasification, pyrolysis, and torrefaction. The wet processes include anaerobic digestion and high-temperature liquefaction.

It should be noted that the price for electricity and natural gas have significant effect on the economics of energy generation and recovery.

4.1 Gasification

The gasification process is the high-temperature and high-pressure conversion of dried sludge into "synthesis" gas (syngas for short) and an ash residue. With gasification, the oxygen is kept at low concentrations to improve the yield of syngas. There has been interest in sludge gasification (U.S. Environmental Protection Agency, 2012). Gasification can provide a clean and easily manageable process, with the possibility of net energy gains. However, the variability and lack of information on commercial-scale systems makes it difficult to evaluate the viability of sludge gasification.

4.2 Pyrolysis

Pyrolysis is the partial decomposition of organic material at high temperatures (typically 450 to 650 °C) in an oxygen-deprived environment, similar to gasification, but curtailed to prevent full conversion to gas. The end products of this process are solid char, liquid oils, and gas. The biochar may be used as a soil amendment. The past decade has seen renewed interest in this technology as an alternative method of sludge disposal combined with energy recovery. Work performed by Agar et al. (2018) demonstrated that pyrolysis char and gas yields showed potential for use as fuels. McNamara et al. (2016) presented the results of bench-scale tests on heat-dried biosolids indicating temperatures up to 500 °C reduced the fraction of biochar, at which point gas yield increased. Pyrolysis also appears to be able to significantly reduce emerging contaminants of concern. Hoffman et al. (2016), described how pyrolysis of anaerobically digested solids at 500 °F (260 °C) resulted in more than 95% removal of estrogenic compounds.

4.3 Torrefaction

Torrefaction is a low-temperature version of pyrolysis, with reduced production of smaller char and incomplete conversion of the offgases to a stable syngas end product (Basu, 2013). Torrefied wastewater biosolids are considered a better fuel than dried biosolids; however, the incomplete conversion of the gases tends to product significant tars and clogging and fires are common with pilot projects.

4.4 High-Temperature Liquefaction

Hydrothermal liquefaction (HTL), is a thermochemical depolymerization process that uses moderate temperature and high pressure to convert wet biomass into biocrude oil and sidestream chemicals. The biocrude appears to have significant ease of use as a petroleum feedstock. However, HTL also produces a high chemical oxygen demand wastewater and this flow stream has been found difficult to treat.

4.5 Heat Recovery

Numerous industrial processes generate waste heat and the close proximity to the typical WRRF facility has driven efforts to capture these heat sources. Waste heat can be captured and repurposed via the following approaches:

- Enhancing industrial processes such as drying or space dehumidification;
- Modifying material properties via warming (e.g., grease or digesters) or preheating combustion air;
- Utilization within steam systems; and
- Direct electrical generation.

Waste heat recovery can make economic sense when included in initial design and construction of a facility (Hawley & Fenner, 2012). Unfortunately, in many instances, retrofits are not financially viable because of high capital costs, with a minimum viable temperature ranging approximately 250 °C (Ling-Chin et al., 2019).

Heat recovery from influent wastewater has been evaluated, but it is unknown whether there are any full-scale implementations of this approach.

4.6 Kinetic Energy Recovery

Kinetic energy recovery systems generate electricity from either moving flow streams or pressure sources within a water utility, where the system has excess energy than what is needed for the operation of the system (New York State Energy Research and Development Authority, 2011). Typical applications include energy dissipation structures within gravity flow systems and pressure relief structures within pressure systems. Other opportunities have been evaluated at the final effluent outfall points but did not go beyond a desk-top evaluation because of concerns about maintaining suitable oxygen levels in the treated water.

5.0 NEW TREATMENT APPROACHES

5.1 Carbon Redirection

Carbon redirection is the process of redirecting substrate toward energy recovery and away from oxidation, and thus both increasing the amount of available carbon and decreasing aeration demands. This is achieved through chemically enhanced primary treatment (CEPT) (Crawford, 2010). With CEPT, one issue to keep in mind is the density of the floc. Polymers can entrain air bubbles and inhibit proper settling. A combination of iron-based coagulant combined with anionic polymer tends to work most effectively and jar testing is recommended.

5.2 Modified Biological Treatment

Emerging biological treatment technologies are generally focused on two main criteria—first, decoupling the solids residence time and the hydraulic residence time, and second, maintaining treatment despite variations in load and flow. Two examples of these are the moving bed biofilm reactor (MBBR) and the membrane aerated biofilm reactor (MABR). An MBBR is a hybrid of a conventional activated sludge system and a fixed-film system, in which floating plastic pieces are used to provide surface attachment points for biological growth. Aeration is provided via conventional means and a sieve or strainer is used to retain the plastic pieces in the aeration chamber (Ødegaard et al., 1994).

Conversely, in an MABR, aeration is provided via thin membranes in which bacteria are allowed to grow, using oxygen that diffuses through the membranes. These systems have shown considerable reduction in required aeration energy (Casey et al., 1999).

5.3 Algae

Algae are aqueous-phase photosynthetic organisms that consume carbon dioxide and release oxygen. When integrated with aerobic treatment, algae can reduce the demand for external oxygen, thus reducing aeration requirements. Much of the commercialization and excitement around algae's role in wastewater treatment has been focused on nutrient removal. Research conducted at The University of Akron discovered a different type of algae and purpose within wastewater treatment (Li et al., 2016; Li and Ju, 2018; Xiao and Ju, 2018a, 2018b). Their research suggests adding a waste activated sludge pretreatment step using phagotrophic algae before sludge digestion can have significant effect on plant energy consumption. When taken a step further, the high lipid algae can be harvested and sold for further processing.

The algae also have energy generation potential through resource recovery process. Incorporating algae as a nutrient removal tool in wastewater treatment may have several advantages:

- Tertiary nutrient removal,
- Algae biomass can be used as a feedstock for energy generation processes,
- Use of carbon dioxide for algal growth can reduce facility carbon footprint, and
- Inherent generation of oxygen can reduce aeration requirements.

5.4 Source Separation—Urine Capture

Although making up only a small percentage of the volume of flow, urine is responsible for up to 80% of the nitrogen and 50% of phosphorous loads in domestic wastewater flows (Jimenez et al., 2015). Nitrogen enters the facility in the form of ammonia, which requires 4.6 kg of oxygen for each kilogram of ammonia to be converted to nitrate. Nitrification can require as much aeration energy as needed for organics removal. Because aeration is a major demand for electricity, reducing ammonia content could have a significant effect on the operating cost of a facility.

Urine peaks generally occur in the morning and taper off as the day goes on. These peaks that last just a fraction of the day have driven the need for aeration systems to be sized larger for the nitrification and denitrification process.

The idea of conserving energy by keeping urine out of the domestic wastewater flows using source separating toilets has been around since the 1990s. Rather than treat the urine at the WRRF, it is diverted and can be reclaimed as a fertilizer source. Special toilets can capture urine and keep it from entering the sanitary sewer system with the rest of household domestic wastewater. Elimination of the ammonia peaks in morning flows can allow engineers to design smaller aeration systems that treat a simpler, more consistent wastewater stream.

Swedish technology for source separation toilets has existed since the 1990s. The technology is promising for rural areas and places that do not currently have infrastructure in place. There may be significant financial and operational challenges to implement this system in already developed areas.

5.5 Dry Digestion

Dry digestion is the digestion of source-separated organics operated at high total solids (15% to 20%) to generate renewable natural gas. Technology vendors claim this system can offer energy recovery along with reusable digestate as soil conditioner. Some of the manufacturers offering these designs include Zero Waste Energy, BioFerm Energy Systems, Hitachi Innovation (Kompogas), Dranco, and Valorga.

5.6 Artificial Intelligence and Operational Guidance

Advanced analytical tools can improve the energy and operational efficiency of wastewater treatment, especially when handling fluctuating load and flow. However, the full use of these analytical tools is dependent on accurate and robust sensor deployment, combined with a seamless data repository (Norton et al., 2019).

Further, mechanistic modeling faces three major challenges:

1. Data originates from physical, chemical, and biological processes. Many variables are strongly coupled, and the data are noisy, uncertain, and incomplete.
2. The process has a temporal nature, e.g., is time-based.
3. The relationships are nonlinear and dynamic.

Wastewater treatment has opportunity for artificial intelligence guidance because of the large amounts of easily available data, complex biological processes, and significant assets to optimize (Wei, 2013). Wei describes the greatest need as predictive and prescriptive solutions that forecast the future and offer control guidance, in a process that involves data acquisition, relationship modeling, decision support, dynamic control, and system optimization.

6.0 SUMMARY AND CONCLUSIONS

Like other industries, the water resource recovery industry is taking a serious look at strategies to improve energy management. Besides energy advantages, other benefits of energy management may include enhanced nutrient recovery and reuse, reduced carbon footprint, and potential reduction in WRRF operating costs.

7.0 REFERENCES

Agar, D. A., Kwapinska, M., & Leahy, J. J. (2018). Pyrolysis of wastewater sludge and composted organic fines from municipal solid waste: Laboratory reactor characterisation and product distribution. *Environmental Science and Pollution Research, 25*, 35874–35882. https://doi.org/10.1007/s11356-018-1463-y

Basu, P. (2013). *Biomass gasification, pyrolysis and torrefaction* (2nd ed.). Academic Press. https://doi.org/10.1016/B978-0-12-396488-5.00014-9

Casey, E., Glennon, B., & Hamer, G. (1999). Review of membrane aerated biofilm reactors. *Resources, Conservation and Recycling, 27*(1–2), 203–215. https://doi.org/10.1016/S0921-3449(99)00007-5

Crawford, G. (2010). *Technology roadmap for sustainable wastewater treatment plants in a carbon-constrained world* (Vol. 9). IWA Publishing. https://doi.org/10.2166/9781780403366

Hawley, C., & Fenner, R. (2012). The potential for thermal energy recovery from wastewater treatment works in southern England. *Journal of Water and Climate Change, 3*, 287. DOI: 10.2166/wcc.2012.013

Hoffman, T. C., Zitomer, D. H., & McNamara, P. J. (2016). Pyrolysis of wastewater biosolids significantly reduces estrogenicity. *Journal of Hazardous Materials, 317*, 579–584.

Jimenez, J., Bott, C., Love, N., & Bratby, J. (2015). Source separation of urine as an alternative solution to nutrient management in biological nutrient removal treatment plants. *Water Environment Research, 87*(12), 2120–2129.

Li, C., & Ju, L.-K. K. (2018). Enhancement of resource recovery and sludge digestion by cultivation of phagotrophic algae with alkali-pretreated waste activated sludge and waste ketchup. *Process Safety and Environmental Protection, 113,* 233–241.

Li, C., Xiao, S., & Ju, L.-K. K. (2016). Cultivation of phagotrophic algae with waste activated sludge as a fast approach to reclaim waste organics. *Water Research, 91,* 195–202.

Ling-Chin, J., Bao, H., Ma, Z., Taylor, W., & Roskilly, A. (2019). State-of-the-art technologies on low-grade heat recovery and utilization in industry. In I. H. Al-Bahadly (Ed.), *Energy conversion: Current technologies and future trends.* IntechOpen. https://doi.org/10.5772/intechopen.78701

McNamara, P. J., Koch, J., Liu, Z., & Zitomer, D. (2016, September). Pyrolysis of dried wastewater biosolids can be energy positive. *Water Environment Research, 88*(9), 804–810.

Norton, J. W., Wood, B., Saparia, B., Jin, S., & Radtke, L. (2019). A journey into smart(er) waters: The Great Lakes Water Authority's experience and approach to smart water systems. *Advances in Water Research, 29*(4), 11–15.

New York State Energy Research and Development Authority. (2011). *Hydropower from wastewater* (Report No. 10879).

Ødegaard, H., Rusten, B., & Westrum, T. (1994). A new moving bed biofilm reactor—Applications and results. *Water Science and Technology, 29*(10–11), 157–165. https://doi.org/10.2166/wst.1994.0757

Uddin, S., Shareef, H., Mohamed, A., Hannan, M. A., & Mohamed, K. (2011). LEDs as energy efficient lighting systems: A detail review. *2011 IEEE Student Conference on Research and Development, Cyberjaya,* pp. 468–472.

U.S. Environmental Protection Agency [Prepared by Greenhouse Gas Technology Center]. (2012, June). *Technology assessment report: Aqueous sludge gasification technologies* (Version 1.5; EPA/600/R-12/540).

Wang, S., Liu, Y., Li, P., Wang, Y., Yang, J., & Zhang, W. (2020). Micronanobubble aeration promotes senescence of submerged macrophytes with low total antioxidant capacity in urban landscape water. *Environment Science: Water Research & Technology, 6,* 523–531. https://doi.org/10.1039/C9EW00929A

Wei, X. (2013). Modeling and optimization of wastewater treatment process with a data-driven approach [PhD (Doctor of Philosophy) thesis]. University of Iowa.

Xiao, S., & Ju, L.-K. K. (2018a). Energy-efficient ultrasonic release of bacteria and particulates to facilitate ingestion by phagotrophic algae for waste sludge treatment and algal biomass and lipid production. *Chemosphere, 209,* 588–598.

Xiao, S., & Ju, L.-K. K. (2018b). Phagotrophic microalgae production from waste activated sludge under non-sterile conditions. *Water Research, 145,* 190–197.

Xiao, W., & Xu, G. (2020). Mass transfer of nanobubble aeration and its effect on biofilm growth: Microbial activity and structural properties. *Science of the Total Environment, 703,* 134976. https://doi.org/10.1016/j.scitotenv.2019.134976

Appendix A

Equations for Converting from U.S. Customary Units to International Standard Units

atm $\times$ 101.3 = kPa
Btu $\times$ 1.055 = kJ
Btu/cu ft $\times$ 37.26 = kJ/m^3
Btu/gal $\times$ 278.7 = kJ/m^3
Btu/hr $\times$ 0.293 1 = W
Btu/lb $\times$ 2.326 = kJ/kg
cfm $\times$ (4.719×10^{-4}) = m^3/s
cfm/ft $\times$ 1.549 = L/m·s
cfs $\times$ (2.832×10^{-2}) = m^3/s
cu ft $\times$ (2.832×10^{-2}) = m^3
cu ft/lb $\times$ (6.243×10^{-2}) = m^3/kg
(°F − 32) 0.555 6 = °C
gal $\times$ (3.785×10^{-3}) = m^3
gal $\times$ 3.785 = L
gph $\times$ (1.051×10^{-6}) = m^3/s
gpm $\times$ (6.308×10^{-5}) = m^3/s
in. $\times$ 25.40 = mm
in. $\times$ (2.540×10^{-2}) = m
ft $\times$ 0.304 8 = m
ft-lb $\times$ 1.356 = N · m
ft-lb/min $\times$ (2.259×10^{-2}) = W
ft-lb/sec $\times$ 1.355 = W
ft/sec $\times$ 0.304 8 = m/s
gal $\times$ (3.785×10^{-3}) = m^3
gpm $\times$ (6.308×10^{-5}) = m^3/s
gpm $\times$ 5.451 = m^3/d
hp $\times$ 745.7 = W
hp-hr $\times$ 2.685 = MJ
hp/mil. gal $\times$ 0.197 0 = W/m^3
in. $\times$ (2.540×10^{-2}) = m
in. Hg $\times$ 3.377 = kPa
kWh = 3.600 MJ
kWh/d $\times$ 41.67 = W
kWh/lb $\times$ (7.936×10^{-3}) = MJ/kg
kWh/mil. gal $\times$ 951.1 = J/m^3
lb $\times$ 0.453 6 = kg
lb/cu ft $\times$ 16.02 = kg/m^3
lb/d $\times$ 5.250 = mg/s
lb/lb $\times$ 1000 = g/kg
mgd $\times$ (4.383×10^{-2}) = m^3/s
mgd $\times$ (3.785×10^{3}) = m^3/d
mile $\times$ 1.609 = km
psi $\times$ 6895 = Pa
scfm $\times$ (4.719×10^{-4}) = m^3/s
sq ft $\times$ (9.290×10^{-2}) = m^2

Appendix B

Energy Tables

Tables in Appendix B—B.1 to B.4—are used to establish benchmarks for water resource recovery facilities (WRRFs). Each table represents a different configuration of the treatment process:

- trickling filter (without nitrification)
- conventional activated sludge (without nitrification)
- conventional activated sludge (without nitrification) and with sand filtration
- activated sludge, single stage nitrification, and sand filtration

These tables have been used by practicing environmental engineers at WRRFs since the mid-1990s. They should provide a reasonable energy benchmark for many different types of WRRF.

1. The table that best matches a specific WRRF is used as the base for the analysis.
2. The columns that best match the average daily flow of the facility are selected.
3. The tables are then adjusted by subtracting or addition of unit processes or operations that were not included in the original table. Values may be estimated.
4. The values of power consumption are interpolated to arrive at the best value for each unit process or operation.
5. The values are then summed to establish the benchmark for the facility.

Updates to the tables were provided in this edition by adding Table B.0, which includes values for additional unit processes. Formatting was adjusted to add subheadings to groups of unit processes.

TABLE B.0. Electricity requirements for trickling filter water resource recovery facilities

The following information is used to "adjust" the values in Tables B.1, B.2, B.3, and B.4 based on unit processes specific to the site.

	Electricity used, kWh/d[a] (except where noted)					
Item	**4-ML/d**	**20-ML/d**	**40-ML/d**	**80-ML/d**	**190-ML/d**	**380-ML/d**
	1	**5**	**10**	**20**	**50**	**100**
	1-mgd[b]	**5-mgd**	**10-mgd**	**20-mgd**	**50-mgd**	**100-mgd**
Grit removal						
Tray Style	15	25	33	56	134	267
Vortex Style	21	31	40	69	159	317
Primary Solids Removal						
Cloth Media Filter	16	98	196	359	839	1595
Biological Treatment						
Bio-Tower with Plastic Media	120	590	1160	2340	5320	n.a.
Sequencing Batch Reactors (SBR) w. Nitrification	1269	6344	12 412	24 825	n.a.	n.a.
Membrane Biological Reactors (MBR) w. Nitrification	1850	9220	18 430	36 710	91 550	182 960
Moving Bed Biofilm Reactor (MBBR) w. Nitrification	1211	2596	5078	10 157	24 590	49 179
Advanced Secondary Treatment						
Disc Filter	3	16	33	68	126	326

Disinfection						
Ultraviolet Disinfection[c]						
Table B.1	123	614	1229	2458	6144	12 288
Table B.2	92	461	922	1843	4608	9216
Table B.3 and Table B.4	77	384	768	1536	3840	7680
Thickening (Sludge Feed Pumps, Polymer Feed, Thickener, and Thickened Sludge Pump)						
Gravity Belt (or Drum) Thickener[d]						
Table B.1 and Table B.2	n.a.[e]	n.a.	288	434	873	1604
Table B.3 and Table B.4	n.a.	n.a.	343	517	1039	1909
Thickening Centrifugec						
Table B.1 and Table B.2	n.a.	n.a.	526	1220	3467	8604
Table B.3 and Table B.4	n.a.	n.a.	627	1453	4126	10 240
Dewatering[d]						
Centrifuge						
Table B.1 and Table B.2	n.a.	460	1030	1485	2832	5141
Table B.3 and Table B.4	n.a.	547	1226	1769	3370	6118
Screw Press						
Table B.1 and Table B.2	n.a.	125	250	376	757	1390
Table B.3 and Table B.4	n.a.	148	297	448	900	1654
Facility Utility Water	12	66	140	292	803	1644

[a]UVT = 65%, TSSmax = 30 mg/Ll, 30 day avg—200 fecal coliform/100 mL and 126 E. coli/100 mL
[b]Includes polymer feed, sludge feed, booster pump, and thickened sludge pumping
[c]Sludge feed, centrifuge, and thickened sludge pumping
[d]Sludge feed pumps, polymer feed, press, pump, or conveyor
[e]Not applicable

TABLE B.1. Electricity requirements for trickling filter water resource recovery facilities

Item	Electricity used, kWh/d[a] (except where noted)					
	4-ML/d	20-ML/d	40-ML/d	80-ML/d	190-ML/d	380-ML/d
	l-mgd[b]	5-mgd	10-mgd	20-mgd	50-mgd	100-mgd
Wastewater Pumping	171	716	1402	2559	6030	11 818
Screens	2	2	2	3	6	11
Grit Removal (Aerated grit)	49	87	134	250	600	1200
Primary Solids Removal (Primary clarifiers)	15	78	155	310	776	1551
Biological Treatment						
Trickling filters[c] (Rock media filter)	352	1319	2528	4686	11 551	22 826
Biological Nitrification	n.a.[d]	n.a.	n.a.	n.a.	n.a.	n.a.
Secondary Clarifiers	15	78	155	310	776	1551
Return Activated Sludge Pumping (with 25% recycle)	n.a.	n.a.	n.a.	n.a.	n.a.	n.a.
Advanced Secondary Treatment						
Intermediate Pumping	n.a.	n.a.	n.a.	n.a.	n.a.	n.a.
Chemical Addition	n.a.	n.a.	n.a.	n.a.	n.a.	n.a.
Sand Filtration	n.a.	n.a.	n.a.	n.a.	n.a.	n.a.
Disinfection (Chlorination)	1	5	27	53	133	266
Thickening[g]						
Gravity Thickening	6	15	25	37	75	138
Dissolved Air Flotation Thickening	n.a.	n.a.	1805	2918	6257	11 819

Stabilization						
Aerobic Digestion[h]	1000	2000	n.a.	n.a.	n.a.	n.a.
Anaerobic Digestion[i]	n.a.	n.a.	1100	2100	5000	11 000
Dewatering[j]						
Belt Filter Press[k]	n.a.	192	384	579	1164	2139
Lighting and Buildings	200	400	800	1200	2000	3000
Totals	1811	4892	8517	15 005	34 368	67 319
Unit Electricity Use, kWh/mil. gal[e]	1811	978	852	750	687	673
Energy Recovery (from biogas combustion)	n.a.	n.a.	2800	5600	14 000	28 000
Net Consumption[f]	1811	4892	5717	9405	20 368	39 319
Unit Net Electricity Use, kWh/mil. gal	1811	978	572	470	407	393

[a]kWh/d × 41.67 = W
[b]mgd × $(4.383 \times 10^{-2}) = m^3/s$
[c]With recirculation pumping
[d]Not applicable
[e]kWh/mil. gal × 951.1 = J/m^3
[f]Total less electricity from energy recovery
[g]Sludge feed pumps, polymer feed, thickener, and thickened sludge pump
[h]Sludge transfer pump, blower, diffusers
[i]Mixers, sludge transfer pump, heating
[j]Sludge feed pumps, polymer feed, press, pump, or conveyor
[k]Sludge feed pumps, polymer feed, thickener, and thickened sludge pump

TABLE B.2. Electricity requirements for activated-sludge water resource recovery facility

	Electricity used, kWh/d[a] (except where noted)					
Item	4-ML/d	20-ML/d	40-ML/d	80-ML/d	190-ML/d	380-ML/d
	1-mgd[b]	5-mgd	10-mgd	20-mgd	50-mgd	100-mgd
Wastewater Pumping	171	716	1402	2559	6030	11 818
Screens	2	2	2	3	6	11
Grit Removal (Aerated grit)	49	87	134	250	600	1200
Primary Solids Removal (Primary clarifiers)	15	78	155	310	776	1551
Biological Treatment (BOD_5 = 150 mg/L)						
Aeration[f]	532	2660	5320	10 640	26 600	53 200
Biological Nitrification	n.a.[c]	n.a.	n.a.	n.a.	n.a.	n.a.
Secondary Clarifiers	15	78	155	310	776	1551
Return Sludge Pumping (with 25% recycle)	45	213	423	724	1627	3131
Advanced Secondary Treatment						
Intermediate Pumping	n.a.	n.a.	n.a.	n.a.	n.a.	n.a.
Chemical Addition	n.a.	n.a.	n.a.	n.a.	n.a.	n.a.
Sand Filtration	n.a.	n.a.	n.a.	n.a.	n.a.	n.a.
Disinfection (Chlorination)	1	5	27	53	133	266
Thickening[g]						
Gravity Thickening	6	15	25	37	75	138
Dissolved Air Flotation Thickening	n.a.	n.a.	1805	2918	6257	11 819

Stabilization						
Aerobic Digestion[h]	1200	2400	n.a.	n.a.	n.a.	n.a.
Anaerobic Digestion[i]	n.a.	n.a.	1400	2700	6500	13 000
Dewatering[j]						
Belt Filter Press (incl. booster pump)	n.a.	192	384	579	1164	2139
Lighting and Buildings	200	400	800	1200	2000	3000
Totals	2236	6846	12 032	22 283	52 544	102 824
Unit Electricity Use, kWh/mil. gal[d]	2236	1369	1203	1114	1051	1028
Energy Recovery (from biogas combustion)	n.a.	n.a.	3500	7000	17 500	35 000
Net Consumption[e]	2236	6846	8532	15 283	35 044	67 824
Unit Net Electricity Use, kWh/mil. gal	2236	1369	853	764	701	678

[a]kWh/d × 41.67 = W

[b]mgd × (4.383×10^{-2}) = m^3/s

[c]Not applicable

[d]kWh/mil. gal × 951.1 = J/m^3

[e]Total less electricity from energy recovery

[f]Fine pore diffused air, BOD_5 = 150 mg/L, AOR:SOR = 0.35, 14 ft submergence

[g]Sludge feed pumps, polymer feed, thickener, and thickened sludge pump

[h]Sludge transfer pump, blower, biffusers

[i]Mixers, sludge transfer pump, peating

[j]Sludge feed pumps, polymer feed, press, pump, or conveyor

TABLE B.3. Electricity requirements for advanced water resource recovery facilities without nitrification

	Electricity use, kWh/d[a] (except where noted)					
Item	4-ML/d	20-ML/d	40-ML/d	80-ML/d	190-ML/d	380-ML/d
	1-mgd[b]	5-mgd	10-mgd	20-mgd	50-mgd	100-mgd
Wastewater pumping	171	716	1402	2559	6030	11 818
Screens	2	2	2	3	6	11
Grit Removal (Aerated grit)	49	87	134	250	600	1200
Primary Solids Removal (Primary clarifiers)	15	78	155	310	776	1551
Biological Treatment (BOD_5 = 150 mg/L)						
Aeration[f]	532	2660	5320	10 640	26 600	53 200
Biological Nitrification	n.a.[c]	n.a.	n.a.	n.a.	n.a.	n.a.
Secondary Clarifiers	15	78	155	310	776	1551
Return Sludge Pumping (with 25% recycle)	45	213	423	724	1627	3131
Advanced Secondary Treatment						
Intermediate or Filter Feed Pumping	143	445	822	1645	3440	6712
Chemical Addition	80	290	552	954	2187	4159
Sand Filtration	137	247	385	709	1679	3295
Disinfection (Chlorination)	1	5	27	53	133	266
Thickening[g]						
Gravity Thickening	6	15	25	37	75	138
Dissolved Air Flotation Thickening	n.a.	n.a.	2022	3268	7008	13 237

Stabilization						
Aerobic Digestion[h]	1200	2400	n.a.	n.a.	n.a.	n.a.
Anaerobic Digestion[i]	n.a.	n.a.	1400	2700	6500	13 000
Dewatering[j]						
Belt Filter Press (incl. booster pump)	n.a.	228	457	689	1385	2545
Lighting and Buildings	200	400	800	1200	2000	3000
Totals	2596	7864	14 081	26 051	60 822	118 814
Unit Electricity Use, kWh/mil. gal[d]	2596	1573	1408	1303	1216	1188
Energy Recovery (from biogas combustion)	n.a.	n.a.	3500	7000	17 500	35 000
Net Consumption[e]	2596	7864	10 581	19 051	43 322	83 814
Unit Net Electricity Use, kWh/mil. gal	2596	1573	1058	953	866	838

[a]kWh/d × 41.67 = W

[b]mgd x (4.383 × 10^{-2}) = m^3/s

[c]Not applicable

[d]kWh/mil. gal × 951.1 = J/m^3

[e]Total less electricity from energy recovery

[f]Fine pore diffused air, BOD_5 = 150 mg/L, AOR:SOR = 0.35, 14 ft submergence

[g]Sludge feed pumps, polymer feed, thickener, and thickened sludge pump

[h]Sludge transfer pump, blower, diffusers

[i]Mixers, sludge transfer pump, heating

[j]Sludge feed pumps, polymer feed, press, pump, or conveyor

TABLE B.4. Electricity requirements for advanced water resource recovery facility with nitrification

Item	Electricity use, kWh/d[a] (except where noted)					
	4-ML/d	20-ML/d	40-ML/d	80-ML/d	190-ML/d	380-ML/d
	1-mgd[b]	5-mgd	10-mgd	20-mgd	50-mgd	100-mgd
Wastewater Pumping	171	716	1402	2559	6030	11 818
Screens	2	2	2	3	6	11
Grit Removal (Aerated grit)	49	87	134	250	600	1200
Primary Solids Removal (Primary clarifiers)	15	78	155	310	776	1551
Biological Treatment (BOD_5 = 150 mg/L)						
Aeration[f]	532	2660	5320	10 640	26 600	53 200
Biological Nitrification	346	1724	3446	6818	16 936	33 800
Secondary Clarifiers	15	78	155	310	776	1551
Return Sludge Pumping (with 25% recycle)	54	256	508	869	1952	3757
Advanced Secondary Treatment						
Intermediate or Filter Feed Pumping	143	445	822	1645	3440	6712
Chemical Addition	80	290	552	954	2187	4159
Sand Filtration	137	247	385	709	1679	3295
Disinfection (Chlorination)	1	5	27	53	133	266
Thickening[g]						
Gravity Thickening	6	15	25	37	75	138
Dissolved Air Flotation Thickening	n.a.[c]	n.a.	2022	3268	7008	13 237

Stabilization						
Aerobic Digestion[h]	1 200	2 400	n.a.	n.a.	n.a.	n.a.
Anaerobic Digestion[i]	n.a.	n.a.	1700	3200	7800	15 600
Dewatering[j]						
Belt Filter Press (incl. booster pump)	n.a.	228	457	689	1385	2545
Lighting and Buildings	200	400	800	1200	2000	3000
Totals	2951	9631	17 912	33 514	79 383	155 840
Unit Electricity Use, kWh/mil. gal[d]	2951	1926	1791	1676	1588	1558
Energy Recovery (from biogas combustion)	n.a.	n.a.	3500	7000	17 500	35 000
Net Consumption[e]	2951	9631	14 412	26 514	61 883	120 840
Unit Net Electricity Use, kWh/mil. gal	2951	1926	1441	1326	1238	1208

[a]kWh/d x 41.67 = W

[b]mgd × (4.383×10^{-2}) = m^3/s

[c]Not applicable

[d]kWh/mil. gal × 951.1 = J/m^3

[e]Total less electricity from energy recovery

[f]Fine pore diffused air, BOD_5 × 150 mg/L, AOR:SOR × 0.35, 14 ft submergence

[g]Sludge feed pumps, polymer feed, thickener, and thickened sludge pump

[h]Sludge transfer pump, blower, diffusers

[i]Mixers, sludge transfer pump, heating

[j]Sludge feed pumps, polymer feed, press, pump, or conveyor

Appendix C

Electricity Basics

Electricity is a condition in which electrons are displaced from atoms. Because it takes energy to displace the electrons, the displaced electrons represent energy that can do useful work. Therefore, electricity is a form of energy that results from an unbalanced atomic condition. If the electrons are merely displaced, a condition of static electricity exists. If the electrons flow back to the atoms, a condition of current electricity exists during the time of movement. Electricity can be generated in several ways, but some other form of energy is required, and the other energy form must be converted to electricity. The most commonly used methods to generate electricity are magnetic (e.g., generators and alternators), chemical (e.g., batteries), and photocells (e.g., light sensors and solar panels), with magnetic generation being the major source of artificially produced electricity in the world. Fossil fuels, hydropower, geothermal energy, nuclear energy, wind, and tidal energy are used to develop the mechanical energy to magnetically produce electricity.

In any electrical circuit, at least four factors are present: electrical pressure (voltage), electron flow or current (amperage), resistance, and power (wattage). Voltage, current, and resistance are related through Ohm's law in the following equation:

$$E = IR \tag{C.1}$$

Where

E = electrical pressure, volts (V);
I = electrical current, amperes (A); and
R = resistance, ohms.

Ohm's law does not apply to alternating current circuits, which contain inductors or capacitors. Another relationship exists for simple circuit power P, in which

$$P = EI \tag{C.2}$$

Where

P = power, watts (W).

Equipment used in water and wastewater treatment facilities is commonly powered by 480-V three-phase alternating current power. In this case, the equation is modified to include power factor and a multiplier (the square root of 3) to determine power. Power and power factor for three-phase circuits is discussed in detail in Chapter 3.

1.0 ALTERNATING CURRENT VOLTAGE

Alternating current voltage is determined by its effectiveness or the power it produces compared to direct current voltage. Because the voltage of each alternating current starts at zero, increases to maximum, then decreases to zero, alternating current voltage is continually varying and reversing and is never at a constant value. Therefore, an effective value has to be determined. When the voltage of an alternation reaches peak, for example 90 deg, it is known as *peak* or *wave-crest voltage*. The effective voltage, or working voltage, of alternating current is 0.707 times the value of the peak voltage. This value was determined by comparing alternating current voltage with direct current voltage in producing heat in a resistor. In terms of effectiveness, 1 alternating current-effective volt (0.707 of the peak alternating current voltage) is equal to 1 direct current volt. Effective voltage is also known as *root-mean-square (rms) voltage*. All common electrical instruments have readouts that are in effective or rms voltage (e.g., 120 V is the effective voltage and 170 V is the peak voltage of common alternating current).

2.0 VOLTAGE DROP

Voltage is electrical pressure. In any series electrical circuit, the total volts applied to the circuit are entirely used. No pressure returns to the generator. All amperes that leave the generator return, but volts are completely spent in the circuit. If ammeter readings are taken around a circuit, it will be found that the ampere readings are all the same. The volts spent or dropped in a part of a circuit are in direct proportion to the resistance in that part of the circuit.

Copper is the most widely used electrical conductor because of its high electrical transfer efficiency and relatively low cost compared to precious metals, which have higher transfer efficiencies. A voltage drop is present in all circuits because of resistance in the conductor. Resistance increases directly with length of wire and inversely with the diameter of the wire.

Resistance caused by improper wire sizes, poor wire connections, or rough contact points in switches results in a voltage drop of the line voltage, as well as buildup of heat and loss of power. For example, in a simple circuit with a device that requires 100 V and 5 A and has a 1-ohm resistance because of the wire, the line voltage needed will be 105 V because the 1-ohm resistance at 5 A will require 5 V of line pressure to overcome the resistance in the wire. Additionally, the power loss is represented by an equation generated by combining Equations C.1 and C.2, as follows:

$$P = I^2R \tag{C.3}$$

In this example, the power loss is 25 W ($5^2 \times 1$). The 25-W power loss is energy lost in the form of heat dissipated over the length of the wire.

Regardless of the form of the resistance, heat will be produced when electricity flows through it. This is true of connections that may loosen from the constant warming and cooling of electric circuits, which causes expansion and contraction of circuit metals or of contact points in switches, which, in turn, wear from use. Loose connections, worn contact points, or other undesirable resistances can be found using infrared thermography or simple voltage measurements across the connection or switch. Infrared photographs or video cameras make heat visible, with greater temperatures producing brighter images. As discussed previously, circuit resistance results in a voltage drop. A simple voltmeter reading across a connector or switch from time to time will show whether resistance is increasing or at excessive levels.

Index

CPSIA information can be obtained
at www.ICGtesting.com
Printed in the USA
BVHW012035100721
611394BV00027B/63

9 781572 784024